Two Dimensional
Spline
Interpolation
Algorithms

Two Dimensional Spline Interpolation Algorithms

Helmuth Späth
Universität Oldenburg
Oldenburg, Germany

CRC Press
Taylor & Francis Group
Boca Raton London New York

CRC Press is an imprint of the
Taylor & Francis Group, an **informa** business

AN A K PETERS BOOK

First published 1995 by A K Peters, Ltd.

Published 2018 by CRC Press
Taylor & Francis Group
6000 Broken Sound Parkway NW, Suite 300
Boca Raton, FL 33487-2742

© 1995 by Taylor & Francis Group, LLC
CRC Press is an imprint of Taylor & Francis Group, an Informa business

First issued in paperback 2019

No claim to original U.S. Government works

ISBN 13: 978-0-367-44992-6 (pbk)
ISBN 13: 978-1-56881-017-1 (hbk)

Visit the Taylor & Francis Web site at
http://www.taylorandfrancis.com

and the CRC Press Web site at
http://www.crcpress.com

Library of Congress Cataloging-in-Publication Data

Späth, Helmuth.
    [Zweidimensionale Spline-Interpolations-Algorithmen. English]
    Two dimensional spline interpolation algorithms / H. Späth.
      p. cm,
    Includes bibliographic references and index.
    ISBN 1-56881-017-2 :
    1. Spline theory--Data processing. 2. Functions--Data processing.
I. Title.
QA224.S6213 1995   95-38858
  511'.42--dc20   CIP

Contents

Preface   vii

I   Spline Interpolation on Rectangular Grids   **1**

1   Polynomial Interpolation   **3**
    1.1.   Rectangular Grids and Product Interpolation   3
    1.2.   The Lagrange Form of the Bivariate Interpolating Polynomial   5
    1.3.   Polynomial Interpolation on Special Triangular Grids   6

2   Bilinear Spline Interpolation   **13**
    2.1.   Searching a Rectangular Grid   13
    2.2.   Bilinear Interpolation on Rectangles   13

3   Biquadratic Spline Interpolants   **31**
    3.1.   Knots the Same as Nodes   31
    3.2.   Knots Different from Nodes   49
    3.3.   Shape Preservation   68
    3.4.   A Local Quadratic Method of Interpolation   68

4   Bicubic Spline Interpolation   **71**
    4.1.   Bicubic Spline Interpolation on Rectangular Grids   71
    4.2.   Parametric Bicubic Spline Interpolation   91

4.3. Bicubic Hermite Spline Interpolation 93
4.4. Semi-Bicubic Hermite Spline Interpolation 103
4.5. Shape Preservation 106
4.6. Biquadratic Histosplines 108

5 **Birational Spline Interpolants**   **121**
5.1. Birational Spline Interpolants on Rectangular Grids 121
5.2. Birational Histosplines 126

II Spline Interpolation for Arbitrarily Distributed Points

II Spline Interpolation for Arbitrarily Distributed Points   **161**

6 **Global Methods without Triangulation**   **163**
6.1. Existence Problems and Goal Setting 163
6.2. Shepard's Method 165
6.3. Hardy's Multiquadrics 173

7 **Triangulations**   **195**

8 **Linear Spline Interpolants over Triangulations**   **201**

9 **The Approximation of First Partial Derivatives**   **215**

10 **Quadratic Spline Interpolants over Triangulations**   **229**

11 **Cubic Spline Interpolation over Triangulations**   **249**

12 **C^1 Spline Interpolation of Degree Five on Triangulations** **275**

   Postscript   **283**

A **Appendix**   **285**

B **List of Subroutines**   **291**

   Bibliography   **293**

   Index   **303**

that these were chosen predominantly on the basis of good visual representability. Problems with numerous hidden surface patches or with larger numbers of nodes are not easily reproduced on a single page.

As he had done for *One Dimensional Spline Interpolation Algorithms*, Mr. Jörg Maier (Dipl. Math.), now a former colleague at the Department of Mathematics, contributed substantially to the testing, the implementation, and the production of the examples. After he left, Mr. Rüdiger Obst (Dipl. Math.) asssumed an even greater role, especially in Part II, and brought the work to completion. The programs for plotting the computed interpolation surfaces — in the examples one sees a three-dimensional representation of the surface above and its contour lines below — are modifications of one that Mr. Andreas Stark (Dipl. Math.) developed during his thesis work. The text was prepared, with their usual precision, by Mrs. Claus and Mrs. Kuhlmann of the Department of Mathematics.

Oldenburg, May 1990   H. Späth

Preface to the English Edition

A number of typographical errors and small discrepancies have been corrected from the German edition. Most of these were discovered by Prof. Len Bos during the translation, which could not have been carried out in a more congenial manner. Many thanks! The handling of publication matters, in this case by Mrs. Alice Peters, was very supportive and extremely reliable.

H. Späth

Preface

This is the continuation of *One Dimensional Spline Interpolation Algorithms* to two dimensions as mentioned in the postscript to that book. We again take the point of view that the nodes (only in the plane) and the values to be interpolated are fixed ahead of time and that no information on a a possible underlying function is available.

In Part I, where the interpolation nodes are supposed to lie on a rectangular grid, we again give elementary but detailed derivations of all the formulas needed for a computer implementation of the interpolation methods. The greater part of the programs developed in this part of the book had not yet been published; the others can be found in similar form in *Spline Algorithms for the Construction of Smooth Curves and Surfaces* ([101]).

In Part II, where the nodes are allowed to be arbitrarily irregularly distributed, we were not able to do the same. For this choice of material, a large number of the programs (Chapters 7, 8, 9, 11, 12) were already available. Their corresponding mathematical background was, on the one hand too extensive, and on the other, not available in detailed form, and hence we often were (had to be) satisfied with descriptive formulations. Besides some smaller programs, the implementation of a quadratic Powell-Sabin spline interpolation method (Chapter 10) is new. It is taken, with permission of the author, from the thesis of Mr. Oliver Moravec (Dipl. Math.), where the mathematical details also may be found.

With regard to the numerous examples, presented with the intention of giving an idea of how the various algorithms work, it should be mentioned

Part I

Spline Interpolation on Rectangular Grids

1

Polynomial Interpolation

1.1. Rectangular Grids and Product Interpolation

In Part I we will only consider interpolation at all of the grid points of a rectangular grid. This problem arises frequently in applications. Suppose that we are given

$$a = x_1 < x_2 < \cdots < x_n = b \quad (n \geq 2),$$
$$c = y_1 < y_2 < \cdots < y_m = d \quad (m \geq 2). \tag{1.1}$$

The nm points in the xy-plane,

$$P_{ij} = (x_i, y_j), \quad i = 1, \cdots, n, \quad j = 1, \cdots, m, \tag{1.2}$$

are called the *grid points* of the (axes parallel) rectangle,

$$R = [a, b] \times [c, d]. \tag{1.3}$$

This rectangle is subdivided into $(n-1)(m-1)$ *subrectangles*,

$$R_{ij} = \{(x, y) : x_i \leq x \leq x_{i+1}, \, y_j \leq y \leq y_{j+1}\}, \tag{1.4}$$

with vertices,

$$(x_i, y_j), \, (x_{i+1}, y_j), \, (x_{i+1}, y_{j+1}), \, \text{and } (x_i, y_{j+1}). \tag{1.5}$$

The $n + m$ *grid lines* are given by

$$x = x_i, \ i = 1, \cdots, n, \quad \text{and} \quad y = y_j, \ j = 1, \cdots, m. \tag{1.6}$$

In this way, the numbers $x_1, \cdots, x_n, y_1, \cdots, y_m$ with property (1.1), define a *rectangular grid* and fix the arrangement of the subrectangles R_{ij}. (A given skew grid or lattice is always the unique, invertible affine image of a rectangular grid.)

The two-dimensional *interpolation problem* for a given grid is then, for given heights (in the z-direction) u_{ij}, $i = 1, \cdots, n$; $j = 1, \cdots, m$, to find a function $F : R \rightarrow R$, defined on R, such that

$$F(x_i, y_j) = u_{ij}, \quad i = 1, \cdots, n; \ j = 1, \cdots, m. \tag{1.7}$$

Typically, we will postulate F to be of a certain form so that then, finding the interpolant amounts to finding the coefficients of certain basis functions. Thus, the the interpolation nodes coincide with the grid points. (Sometimes we will also allow given values for certain partial derivatives.)

Now if a set $\{L_k^n\}_{k=1,\cdots,n}$ of *cardinal functions*, i.e., $L_k^n : \mathbb{R} \rightarrow \mathbb{R}$ and

$$L_k^n(x_i) = \delta_{ik} = \begin{cases} 1 & \text{if } i = k \\ 0 & \text{if } i \neq k \end{cases}, \tag{1.8}$$

is available, then clearly the *product,*

$$F(x, y) = \sum_{k=1}^{n} \sum_{\ell=1}^{m} u_{kl} L_k^n(x) L_\ell^m(y) =$$

$$(L_1^n(x), \cdots, L_n^n(x)) \begin{pmatrix} u_{11} & u_{12} & \cdots & u_{1,m} \\ u_{21} & u_{22} & \cdots & u_{2,m} \\ \cdot & \cdot & & \cdot \\ \cdot & \cdot & & \cdot \\ u_{n1} & u_{n2} & \cdots & u_{nm} \end{pmatrix} \begin{pmatrix} L_1^m(y) \\ L_2^m(y) \\ \cdot \\ \cdot \\ L_m^m(y) \end{pmatrix}, \tag{1.9}$$

solves the interpolation problem (1.7).

Unfortunately, as we shall see in the examples to follow, simply constructed cardinal functions do not always produce visually pleasing and smooth interpolation surfaces $F = F(x, y)$.

1.2. The Lagrange Form of the Bivariate Interpolating Polynomial

If we take as cardinal functions, L_k^n, the fundamental Lagrange polynomials of univariate polynomial interpolation ([55,100]),

$$L_k^n(x) = L_k^n(x, x_1, \cdots, x_n) = \prod_{\substack{i=1 \\ i \neq k}} \frac{x - x_i}{x_k - x_i}, \qquad (1.10)$$

then (1.9) results in a bivariate polynomial,

$$F(x,y) = \sum_{k=1}^{n} \sum_{\ell=1}^{m} a_{k\ell} x^{k-1} y^{\ell-1} =$$

$$(1, x, x^2, \cdots, x^{n-1}) \begin{pmatrix} a_{11} & a_{12} & \cdots & a_{1m} \\ a_{21} & a_{22} & \cdots & a_{2m} \\ \cdot & \cdot & & \cdot \\ \cdot & \cdot & & \cdot \\ a_{n1} & a_{n2} & \cdots & a_{nm} \end{pmatrix} \begin{pmatrix} 1 \\ y \\ \cdot \\ \cdot \\ y^{m-1} \end{pmatrix}, (1\ 11)$$

in the two variables, x and y. The nm polynomial coefficients $a_{k\ell}$, $k = 1, \cdots, n$; $\ell = 1, \cdots, m$, depend in a complicated way on the cardinal functions. The representation (1.9) with the L_k^n given by (1.10) can also be used directly to evaluate F. This, however, has a relatively high computational expense. It is somewhat more elegant and economical to use the Newton form of the bivariate interpolating polynomial([55]). By no means should the interpolation conditions (1.7) be solved numerically for (1.11), i.e., the linear system,

$$\sum_{k=1}^{n} \sum_{\ell=1}^{m} a_{k\ell} x_i^{k-1} y_j^{\ell-1} = u_{ij}, \quad i = 1, \cdots, n; j = 1, \cdots, m,$$

should never be used to determine the $a_{k\ell}$.

While in this manner quite acceptable interpolation surfaces result for small values of n and m (Figs. 1.1 and 1.2), for larger values of n and m and/or certain configurations of the data, the resulting surfaces have a very wavy appearance. Sometimes it is even difficult to make sense of their surface plots (Figs. 1.3–1.5). (Compare with the results of the methods to follow.)

Here, as throughout, the upper part of the figures show the given grid. The values (1.1) can be read off from the intersections of the grid lines with the axes. The interpolation surfaces also show (thick lines) the projections of the grid lines onto the surface, and hence, taking into account the foreshortening of the z-axis, the heights u_{ij} can also often be read off the graph.

In the lower part of the figures, we give a *contour* (curves of equal function value) plot of the surface. (All figures were produced using the graphics package DISSPLA, which we have available on the Siemens 7890C here at the University of Oldenburg. We used slightly modified versions of the plotting programs of [102].)

We obtain smoother interpolants as well as allow for the possibility of shape preservation (positivity, monotonicity, convexity) in exactly the same way as in the univariate case. We define low degree, piecewise (on each sub-rectangle R_{ij}) bivariate polynomials (or other types of functions) and join them together to form a surface of a certain smoothness. More precisely, we consider

$$F_{ij}(x,y) = \sum_{k=1}^{N} \sum_{\ell=1}^{N} a_{ijk\ell}(x-x_i)^{k-1}(y-y_j)^{\ell-1},$$

$$(x,y) \in R_{ij}, \quad i=1,\cdots,n-1; \, j=1,\cdots,m-1, \qquad (1.12)$$

with $N = 2,3,4$. For $N = 2$, we will join these $(n-1)(m-1)$ functions F_{ij} to be continuous across the boundaries of the rectangles. For $N = 3$, we will join them to be once continuously differentiable, and for $N = 4$ twice continuously differentiable. (Later, we will allow other functions $g_{ik}(x)$ instead of $(x-x_i)^{k-1}$.) The *knots* of such a polynomial spline interpolant are exactly the grid points of R.

1.3.   Polynomial Interpolation on Special Triangular Grids

For arbitrarily distributed interpolation nodes, the bivariate polynomial interpolation problem sometimes has no solution, sometimes has many solutions, and sometimes has a unique solution. Precisely when each of these cases occurs depends on the configuration of the data and the polynomial basis used (see the beginning of Part II). We give an example here of unique solvability, which is of some possible relevance.

If we remove all the grid points P_{ij} with $i+j \geq n+2$ from a rectangular grid with $n = m$, then the remaining $n(n+1)/2$ points P_{ij} with $i+j = 2,\cdots,n+1$, form a *triangular array*. For these, the interpolation problem with bivariate polynomials,

$$F(x,y) = \sum_{k=1}^{n} \sum_{\ell=1}^{n+1-k} b_{k\ell} x^{k-1} y^{\ell-1}, \qquad (1.13)$$

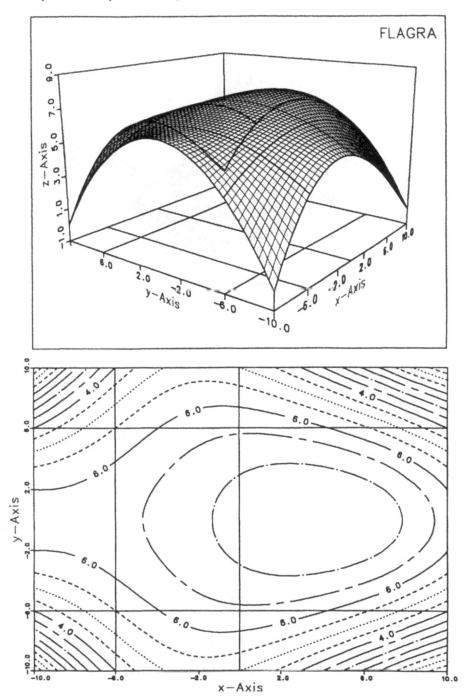

Figure 1.1.

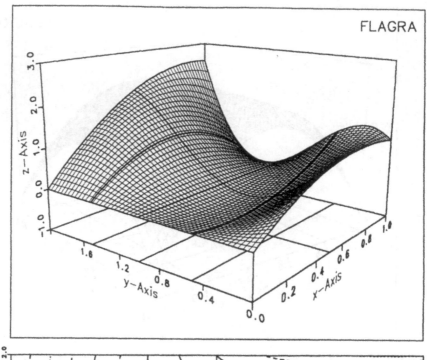

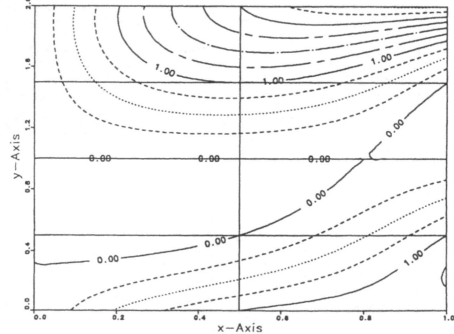

Figure 1.2.

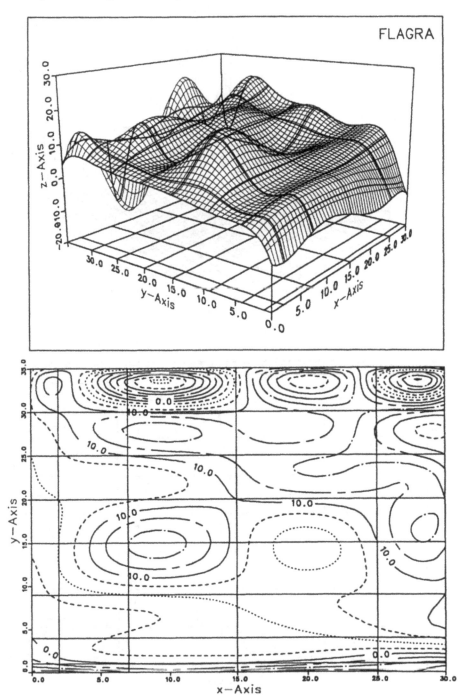

Figure 1.3.

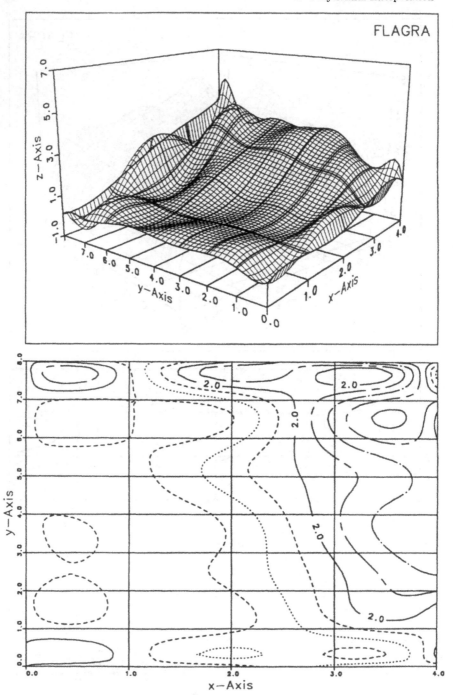

Figure 1.4.

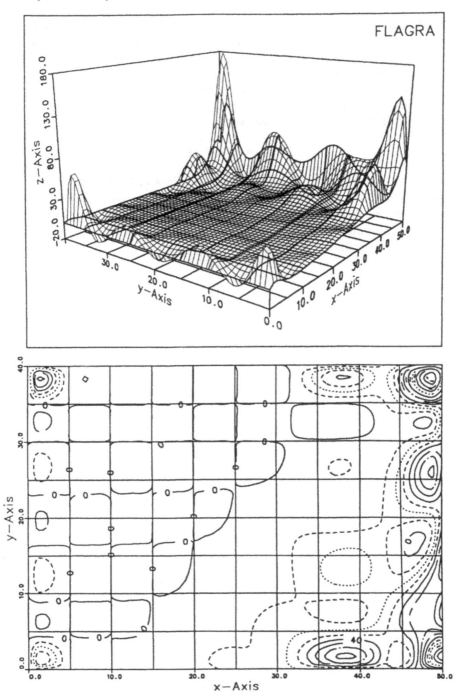

Figure 1.5.

of degree (highest exponent of x or y) $n-1$ is always uniquely solvable ([55]). This fact is also true even if (1.1) does not hold, and hence the resulting array does not look triangular, but as long as the x_i and y_j are both still pairwise distinct.

What does this say when $n=2$? In this case, we are given three points, $(x_1, y_1), (x_1, y_2)$, and (x_2, y_1), with $x_1 \neq x_2$, $y_1 \neq y_2$. Equation (1.13) reduces to the equation of a plane,

$$F(x, y) = b_{11} + b_{12}y + b_{21}x. \qquad (1.14)$$

Clearly, through any three such distinct points in three-space (arbitrary heights u_{11}, u_{12}, and u_{21}), there is exactly one *interpolating plane*. This is even true when the three points are in general position. The only restriction is that the points in $\mathbb{R}^3$ do not lie on a straight line. In that case, there would be an infinite number of interpolating planes.

2

Bilinear Spline Interpolation

2.1.   Searching a Rectangular Grid

Once the $(n-1)(m-1)N^2$ coefficients $a_{ijk\ell}$ of (1.12) have been found, the global function defined piecewise by

$$F|_{R_{ij}} = F_{ij}, \quad i = 1, \cdots, n-1; \, j = 1, \cdots, m-1, \qquad (2.1)$$

has to be evaluated at an arbitrary point $(x, y) \in R$. For this, it is first necessary to find in which of the rectangles R_{ij} the point (x, y) lies.

Just as in the univariate case, this search can be carried out more or less efficiently. The subroutine INTTWO (Figs. 2.1 and 2.2) simply uses the method employed by INTONE ([100]), first in the x direction and then in the y direction. I and J must be initialized in the calling program. Usually, the statement DATA I,J/2*1/ will suffice.

2.2.   Bilinear Interpolation on Rectangles

Interpolation methods for the problem (1.7) are said to be *global* when the evaluation at an arbitrary point $(x, y) \in R$ depends on *all* of the given data (x_i, y_j, u_{ij}), $i = 1, \cdots, n; \, j = 1, \cdots, m$. *Local* methods are those that use only the data (possibly also values for derivatives) of the same rectangle

13

(possibly also the immediately neighboring rectangles) for the evaluation of $F(x,y)$ at $(x,y) \in R_{ij}$. Bilinear interpolation is a local method for a particular rectangle R_{ij}. On each subrectangle R_{ij} we set a bivariate polynomial of the form (1.12) with $N = 2$, i.e.,

$$F_{ij}(x,y) = a_{ij11} + a_{ij21}x + a_{ij12}y + a_{ij22}xy. \tag{2.2}$$

Each of these has four coefficients. F_{ij} is said to be *bilinear*, as it is linear in the y direction for each fixed x and linear in the x direction for each

```
      SUBROUTINE INTTWO(X,N,Y,M,V,W,I,J,IFLAG)
      DIMENSION X(N),Y(M)
      IFLAG=0
      IF(V.LT.X(1).OR.V.GT.X(N).OR.W.LT.Y(1).OR.W.GT.Y(M)) THEN
          IFLAG=3
          RETURN
      END IF
      IF(I.LT.1.OR.I.GE.N.OR.J.LT.1.OR.J.GE.M) THEN
          I=1
          J=1
      END IF
      IF(V.LT.X(I)) GOTO 10
      IF(V.LE.X(I+1)) GOTO 40
      L=N
      GOTO 30
10    L=I
      I=1
20    K=(I+L)/2
      IF(V.LT.X(K)) THEN
          L=K
      ELSE
          I=K
      END IF
30    IF(L.GT.I+1) GOTO 20
40    IF(W.LT.Y(J)) GOTO 50
      IF(W.LE.Y(J+1)) RETURN
      L=M
      GOTO 70
50    L=J
      J=1
60    K=(J+L)/2
      IF(W.LT.Y(K)) THEN
          L=K
      ELSE
          J=K
      END IF
70    IF(L.GT.J+1) GOTO 60
      RETURN
      END
```

Figure 2.1. Program listing of INTTWO.

Calling sequence:

CALL INTTWO(X,N,Y,M,V,W,I,J,IFLAG)

Purpose:
Suppose that we are given $x_1 < \cdots < x_n$, $y_1 < \cdots < y_m$ and that R is the associated rectangular grid. The subroutine localizes the point (v, w) within R, i.e., it determines the index pair (i, j) for which $x_i \leq v \leq x_{i+1}$ and $y_j \leq w \leq y_{j+1}$.

Description of the parameters:

X	ARRAY(N): x values of the rectangular grid R.
N	Number of x values.
Y	ARRAY(N): y values of the rectangular grid R.
M	Number of y values.
V	x-coordinate of the point that is to be localized within R. Restriction: $X(1) \leq V \leq X(N)$.
W	y-coordinate of the point that is to be localized within R. Restriction: $Y(1) \leq W \leq Y(M)$.
I	Input: A value between 1 and $N - 1$. Output: I with $X(I) \leq V \leq X(I+1)$.
J	Input: A value between 1 and $M - 1$. Output: J with $Y(J) \leq W \leq Y(J+1)$.
IFLAG	=0: Normal execution.
	=3: $V<X(1)$, $V>X(N)$, $W<Y(1)$, or $W>Y(M)$.

Figure 2.2. Description of INTTWO.

fixed y. In economics, the x and y terms are interpreted as main effects and the xy term as the interaction effect ([76]). Since on the edges of each rectangle the bilinears defined on either side of the edge restrict to exactly the same linear, these two bilinears join continuously. The result is a C^0 bilinear spline interpolant on all of R.

The coefficients of (2.2) are obtained immediately by taking $n = m = 2$ in (1.9) and replacing (x_1, x_2) by (x_i, x_{i+1}), (y_1, y_2), by (y_j, y_{j+1}), and $(u_{11}, u_{12}, u_{21}, u_{22})$ by $(u_{ij}, u_{i,j+1}, u_{i+1,j}, u_{i+1,j+1})$, and using (1.10), also with $n = 2$. This results in

$$
\begin{aligned}
F_{ij}(x,y) &= \left(\frac{x - x_{i+1}}{x_i - x_{i+1}}, \frac{x - x_i}{x_{i+1} - x_i} \right) \left(\begin{array}{cc} u_{ij} & u_{i,j+1} \\ u_{i+1,j} & u_{i+1,j+1} \end{array} \right) \left(\begin{array}{c} \frac{y - y_{j+1}}{y_j - y_{j+1}} \\ \frac{y - y_j}{y_{j+1} - y_j} \end{array} \right) \\
&= u_{ij} \frac{x - x_{i+1}}{x_i - x_{i+1}} \frac{y - y_{j+1}}{y_j - y_{j+1}} \\
&\quad + u_{i,j+1} \frac{x - x_{i+1}}{x_i - x_{i+1}} \frac{y - y_j}{y_{j+1} - y_j} \\
&\quad + u_{i+1,j} \frac{x - x_i}{x_{i+1} - x_i} \frac{y - y_{j+1}}{y_j - y_{j+1}} \\
&\quad + u_{i+1,j+1} \frac{x - x_i}{x_{i+1} - x_i} \frac{y - y_j}{y_{j+1} - y_j} \\
&= \frac{1}{\Delta x_i \Delta y_j} [u_{ij}(x - x_{i+1})(y - y_{j+1}) - u_{i,j+1}(x - x_{i+1})(y - y_j) \\
&\quad - u_{i+1,j}(x - x_i)(y - y_{j+1}) + u_{i+1,j+1}(x - x_i)(y - y_j)].
\end{aligned}
\tag{2.3}
$$

A form somewhat more efficient for evaluation is obtained by starting with

$$
F_{ij}(x,y) = b_1 + b_2 \frac{x - x_i}{\Delta x_i} + b_3 \frac{y - y_j}{\Delta y_j} + b_4 \frac{(x - x_i)(y - y_j)}{\Delta x_i \Delta y_j},
\tag{2.4}
$$

where we have suppressed the dependence of the coefficients on i and j. The interpolation conditions then give

$$
\begin{aligned}
u_{ij} &= b_1, \\
u_{i,j+1} &= b_1 + b_3, \\
u_{i+1,j} &= b_1 + b_2, \\
u_{i+1,j+1} &= b_1 + b_2 + b_3 + b_4,
\end{aligned}
$$

which may be solved for

$$
\begin{aligned}
b_1 &= u_{ij}, \\
b_2 &= u_{i+1,j} - u_{ij}, \\
b_3 &= u_{i,j+1} - u_{ij}, \\
b_4 &= u_{i+1,j+1} - u_{i,j+1} - u_{i+1,j} + u_{ij}.
\end{aligned}
\tag{2.5}
$$

The subroutine SBILIN (Fig. 2.3) implements C^0 bilinear spline interpolation on R using formulas (2.5) and (2.4) and calling INTTWO.

Several examples are given in Figs. 2.4–2.13. Figures 2.6, 2.7, 2.8, 2.10 and 2.12 correspond to Figs. 1.1–1.5. From the surface and contour plots of the these new figures, it is clear that bilinear functions are by no means linear in the sense of planes. They are only linear for constant x or y, which is also very clear from the plots.

```
      FUNCTION SBILIN(X,N,Y,M,U,NDIM,V,W,IFLAG)
      DIMENSION X(N),Y(M),U(NDIM,M)
      DATA I,J/1,1/
      IFLAG=0
      IF(N.LT.2.OR.M.LT.2) THEN
           IFLAG=1
           RETURN
      END IF
      CALL INTTWO(X,N,Y,M,V,W,I,J,IFLAG)
      IF(IFLAG.NE.0) RETURN
      DV=V-X(I)
      DW=W-Y(J)
      DX=X(I+1)-X(I)
      DY=Y(J+1)-Y(J)
      H1=U(I,J)
      H2=U(I+1,J)
      H3=U(I,J+1)
      H4=(I+1,J+1)
      SBILIN=H1+(H2-H1)/DX*DV+(H3-H1)/DY*DW+
     &        (H4-H3-H2+H1)/(DX*DY)*DV*DW
      RETURN
      END
```

Calling sequence:

CALL SBILIN(X,N,Y,M,U,NDIM,V,W,IFLAG)

Purpose:
Calculation of the function value at the point $(v, w) \in R$ of a bilinear spline interpolant defined on a rectangular grid R.

Description of the parameters:

X,N,Y,M as in INTTWO.

U	ARRAY(NDIM,N): Upon calling must contain the heights that are to be interpolated over the rectangular grid.
NDIM	Maximum first dimension of U. Restriction: NDIM$\geq$N.
V	x-coordinate of the evaluation point.
W	y-coordinate of the evaluation point.
IFLAG	=0: Normal execution.
	=1: N<2 or M<2.

Required subroutine: INTTWO.

Remark: The statement 'DATA I,J/1,1/' has the effect that I and J are both set to 1 at the first call to SBILIN.

Figure 2.3. Program listing of SBILIN and its description.

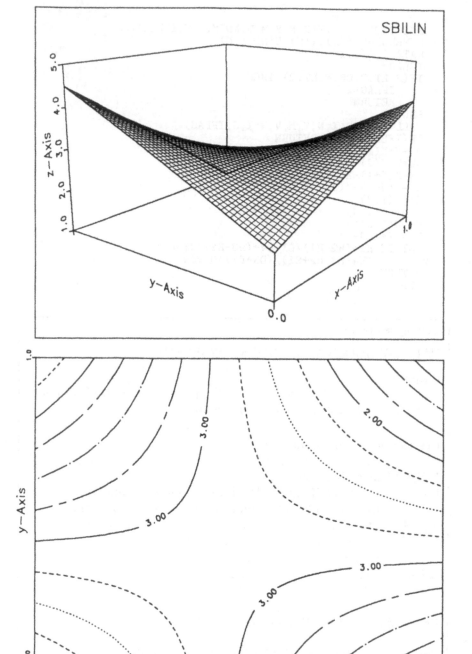

Figure 2.4.

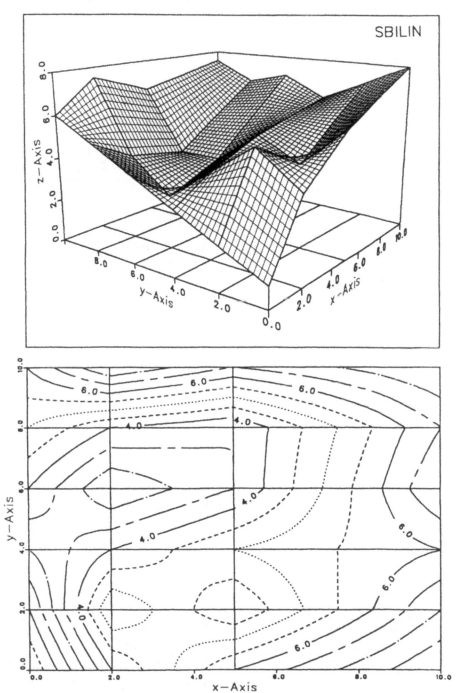

Figure 2.5.

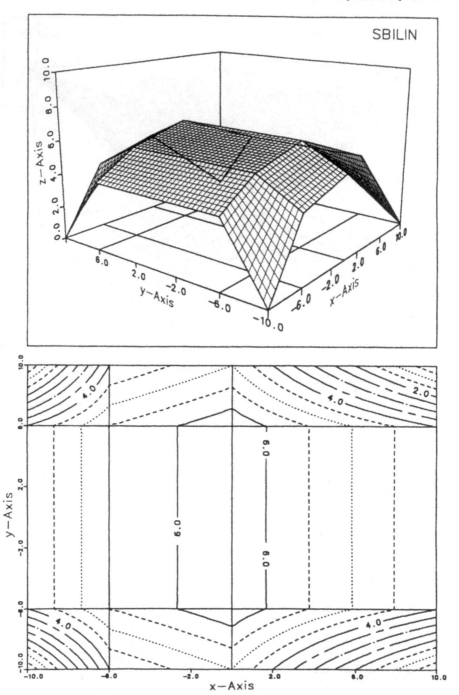

Figure 2.6.

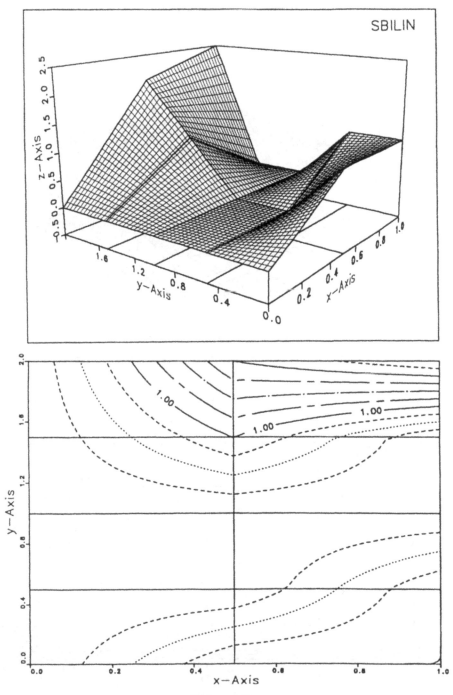

Figure 2.7.

2. Bilinear Spline Interpolation

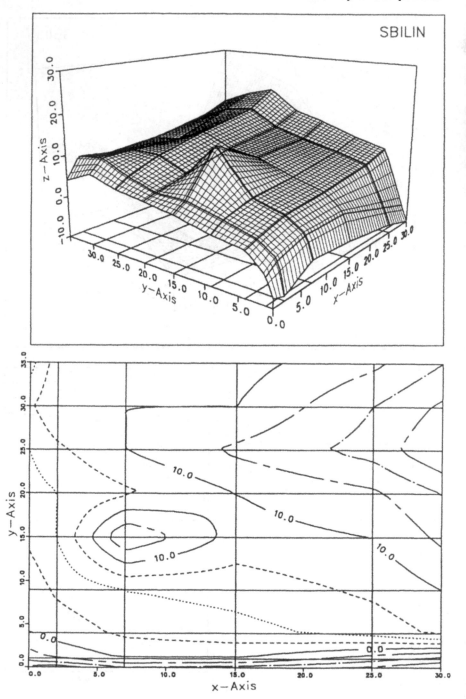

Figure 2.8.

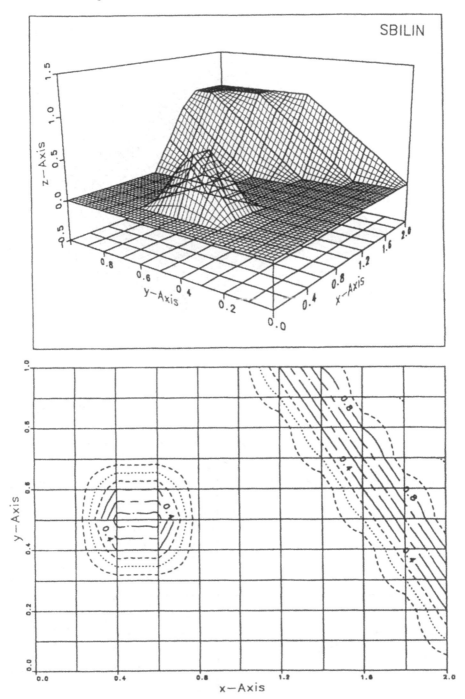

Figure 2.9.

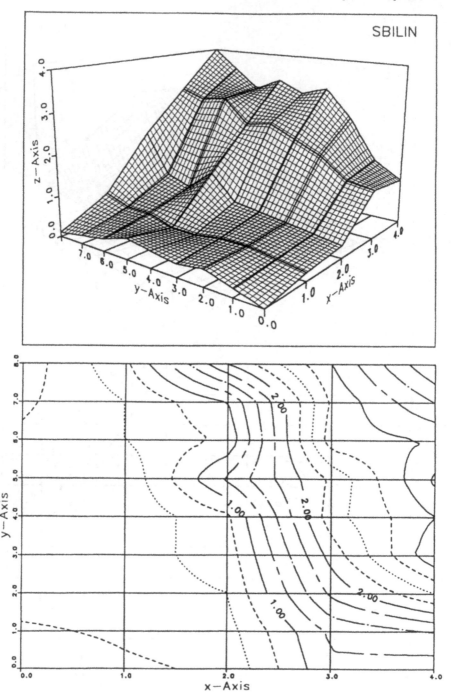

Figure 2.10.

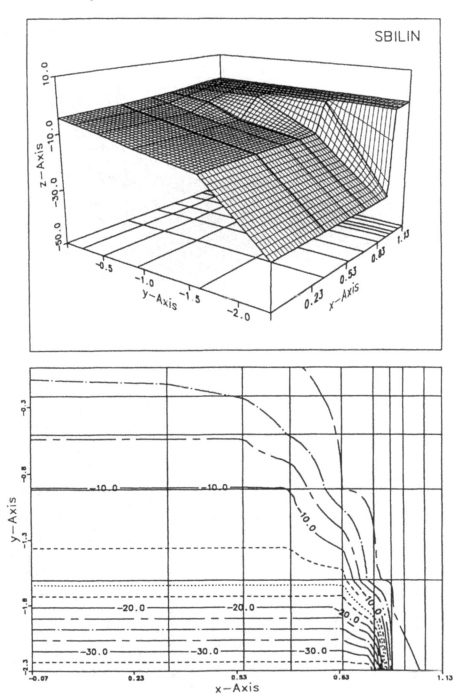

Figure 2.11.

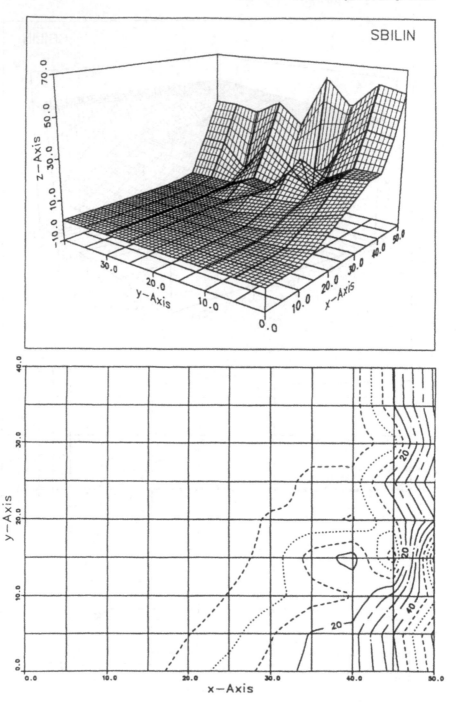

Figure 2.12.

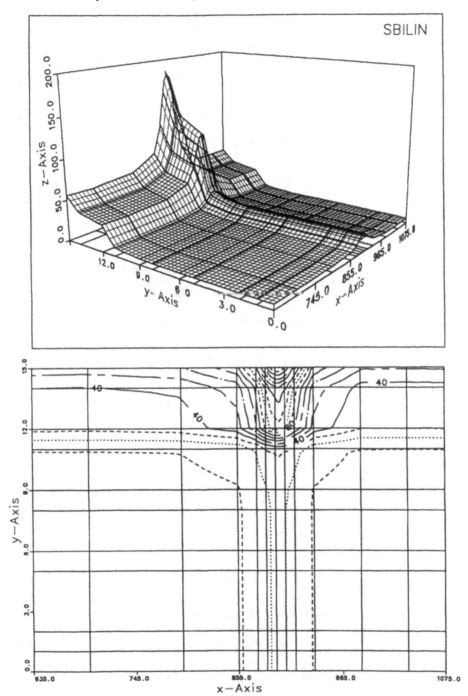

Figure 2.13.

For pedagogical reasons, and since we will be doing something similar for functions (1.12) with $N > 2$, we would like to present a formally somewhat different procedure for computing the coefficients $b_1, \cdots, b_4$ of (2.4). We will now refer again to these more precisely as $a_{ij11}, a_{ij21}, a_{ij12}$, and a_{ij22}. With

$$g_{i1}(x) = 1, \quad g_{i2}(x) = x - x_i, \tag{2.6}$$

(2.4) may be rewritten in the form,

$$
\begin{aligned}
F_{ij}(x, y) &= \sum_{k=1}^{2} \sum_{\ell=1}^{2} a_{ijk\ell} g_{ik}(x) g_{j\ell}(y) \\
&= (g_{i1}(x), g_{i2}(x)) \begin{pmatrix} a_{ij11} & a_{ij12} \\ a_{ij21} & a_{ij22} \end{pmatrix} \begin{pmatrix} g_{j1}(y) \\ g_{j2}(y) \end{pmatrix}. \tag{2.7}
\end{aligned}
$$

(To be more precise, we should write $g_{ik}(x, x_i)$ instead of $g_{ik}(x)$.) The interpolation requirements then give

$$u_{ij} = (1, 0) A \begin{pmatrix} 1 \\ 0 \end{pmatrix},$$

$$u_{i,j+1} = (1, 0) A \begin{pmatrix} 1 \\ \Delta y_j \end{pmatrix},$$

$$u_{i+1,j} = (1, \Delta x_i) A \begin{pmatrix} 1 \\ 0 \end{pmatrix},$$

$$u_{i+1,j+1} = (1, \Delta x_i) A \begin{pmatrix} 1 \\ \Delta y_j \end{pmatrix},$$

where we have set

$$A = \begin{pmatrix} a_{ij11} & a_{ij12} \\ a_{ij21} & a_{ij22} \end{pmatrix}.$$

Together these may be written as

$$\begin{pmatrix} 1 & 0 \\ 1 & \Delta x_i \end{pmatrix} \begin{pmatrix} a_{ij11} & a_{ij12} \\ a_{ij21} & a_{ij22} \end{pmatrix} \begin{pmatrix} 1 & 1 \\ 0 & \Delta y_j \end{pmatrix} = \begin{pmatrix} u_{ij} & u_{i,j+1} \\ u_{i+1,j} & u_{i+1,j+1} \end{pmatrix}. \tag{2.8}$$

Written in terms of matrices, this equation is of the form,

$$V(x_i) A [V(y_j)]^t = U, \tag{2.9}$$

from which it follows that

$$A = [V(x_i)]^{-1} U [V(y_j)]^{-t}, \tag{2.10}$$

so that

$$\begin{pmatrix} a_{ij11} & a_{ij12} \\ a_{ij21} & a_{ij22} \end{pmatrix}$$

$$= \begin{pmatrix} 1 & 0 \\ -\frac{1}{\Delta x_i} & \frac{1}{\Delta x_i} \end{pmatrix} \begin{pmatrix} u_{ij} & u_{i,j+1} \\ u_{i+1,j} & u_{i+1,j+1} \end{pmatrix} \begin{pmatrix} 1 & -\frac{1}{\Delta y_j} \\ 0 & \frac{1}{\Delta y_j} \end{pmatrix}$$

$$= \begin{pmatrix} u_{ij} & \dfrac{u_{i,j+1} - u_{ij}}{\Delta y_j} \\ \dfrac{u_{i+1,j} - u_{ij}}{\Delta x_i} & \dfrac{u_{i+1,j+1} - u_{i,j+1} - u_{i+1,j} + u_{ij}}{\Delta x_i \Delta y_j} \end{pmatrix}.$$

This agrees with the formula for the coefficients of (2.4) given in (2.5). The matrices $V(x_i)$ and $V(y_j)$ are called *connection matrices*.

3

Biquadratic Spline Interpolants

3.1. Knots the Same as Nodes

As we indicated when introducing (1.12), on each subrectangle R_{ij} of our rectangle R we set a biquadratic function $G = F_{ij}$ of the form analogous to (2.7),

$$
G(x, y) = (g_{i1}(x), g_{i2}(x), g_{i3}(x)) \begin{pmatrix} a_{11} & a_{12} & a_{13} \\ a_{21} & a_{22} & a_{23} \\ a_{31} & a_{32} & a_{33} \end{pmatrix} \begin{pmatrix} g_{j1}(y) \\ g_{j2}(y) \\ g_{j3}(y) \end{pmatrix}, \quad (3.1)
$$

where

$$
g_{i1}(x) = 1, \quad g_{i2}(x) = x - x_i, \quad \text{and} \quad g_{i3}(x) = (x - x_i)^2. \quad (3.2)
$$

More precisely, by $g_{ik}(x)$ we actually mean $g_{ik}(x, x_i) = (x - x_i)^{k-1}$, so that $g_{jk}(y)$ should be read as $g_{jk}(y) = (y - y_j)^{k-1}$. (A somewhat different form is chosen in [63]. More general discussions of product interpolation can be found in [22,58], especially the application of B-splines in [22].) We have suppressed the dependence of the nine coefficients $a_{ijk\ell}$, $k, \ell = 1, 2, 3$, on the indices i and j, as for the time being we will only consider a fixed subrectangle R_{ij}, $i = 1, \cdots, n-1$; $j = 1, \cdots, m-1$. Later, we will determine the collective $9(n-1)(m-1)$ coefficients so that the assembled function F

is once continuously differentiable on all of R. This will result in a *global* interpolation method.

Using (3.2), we could also write G in the form,

$$G(x, y) = \sum_{k=1}^{3} \sum_{\ell=1}^{3} a_{ijk\ell}(x - x_i)^{k-1}(y - y_j)^{\ell-1}, \qquad (3.3)$$

instead of (3.1). It is better, however, to use (3.1) to write the partial derivatives in the form,

$$G_x(x, y) = (0, 1, 2(x - x_i))A \begin{pmatrix} 1 \\ y - y_j \\ (y - y_j)^2 \end{pmatrix}, \qquad (3.4)$$

$$G_y(x, y) = (1, x - x_i, (x - x_i)^2)A \begin{pmatrix} 0 \\ 1 \\ 2(y - y_j) \end{pmatrix}, \qquad (3.5)$$

$$G_{xy}(x, y) = (0, 1, 2(x - x_i))A \begin{pmatrix} 0 \\ 1 \\ 2(y - y_j) \end{pmatrix}, \qquad (3.6)$$

where A (actually still dependent on i and j) is the 3×3 matrix of (3.1). In the univariate case, we used the first derivatives at the nodes as unknowns, at least when these agreed with the knots ([100]), and this is basically what we will first do here. Further, we used the first derivative at the left endpoint as boundary condition, either to just specify it or to choose it in a certain sense optimally. In the bivariate case, we will use the values,

$$\begin{aligned} p_{ij} &= G_x(x_i, y_j), \\ p_{i,j+1} &= G_x(x_i, y_{j+1}), \\ q_{ij} &= G_y(x_i, y_j), \\ q_{i+1,j} &= G_y(x_{i+1,j}), \\ r_{ij} &= G_{xy}(x_i, y_j), \end{aligned} \qquad (3.7)$$

as unknowns for subrectangle R_{ij}.

For this choice of unknowns, the desired global function F, defined on all of R, will be uniquely determined and even easy to calculate when, for example, the values,

$$\begin{aligned} p_{1j}, \quad &j = 1, \cdots, m, \\ q_{i1}, \quad &i = 1, \cdots, n, \\ r_{11}, & \end{aligned} \qquad (3.8)$$

for the partial derivatives with respect to x at the grid points on the left edge, for the derivatives with respect to y at the grid points along the bottom edge, and for the mixed partial at the bottom left-hand corner, are either specified or chosen to be in some sense optimal.

The interpolation requirements at the four corners of R_{ij} together with conditions (3.7) for the five derivative values can be written as a linear system of nine equations in nine unknowns, $A = (a_{k\ell})_{k,\ell=1,2,3}$. The inverse of the corresponding coefficient matrix can even be explicitly given ([64]). But here we will follow the more elegant procedure introduced in connection with (2.7) ([101]). If we set

$$
C = \begin{pmatrix}
u_{ij} & q_{ij} & u_{i,j+1} \\
p_{ij} & r_{ij} & p_{i,j+1} \\
u_{i+1,j} & q_{i+1,j} & u_{i+1,j+1}
\end{pmatrix}
\qquad (3.9)
$$

and

$$
\begin{aligned}
V(x_i) &= \begin{pmatrix}
g_{i1}(x_i) & g_{i2}(x_i) & g_{i3}(x_i) \\
g'_{i1}(r_i) & g'_{i2}(x_i) & g'_{i3}(x_i) \\
g_{i1}(x_{i+1}) & g_{i2}(x_{i+1}) & g_{i3}(x_{i+1})
\end{pmatrix} \\
&= \begin{pmatrix}
1 & 0 & 0 \\
0 & 1 & 0 \\
1 & \Delta x_i & \Delta x_i^2
\end{pmatrix},
\end{aligned}
\qquad (3.10)
$$

then from (3.1), (3.4), (3.5), and (3.6), the nine conditions can be written in matrix form as

$$
C = V(x_i)A[V(y_j)]^T.
\qquad (3.11)
$$

This follows easily from the fact that the row vectors of the connection matrix $V(x_i)$ are precisely those, evaluated at x_i and x_{i+1}, that multiply A on the left in (3.1), (3.4), (3.5), and (3.6). Similarly, the row vectors of $V(y_j)$ are those, evaluated at y_j and y_{j+1}, that multiply A on the right. The positions of the elements of C were simply chosen to be consistent with these observations.

The 3×3 matrix $V(x_i)$ is easily inverted. We have

$$
[V(x_i)]^{-1} = \begin{pmatrix}
1 & 0 & 0 \\
0 & 1 & 0 \\
-\frac{1}{\Delta x_i^2} & -\frac{1}{\Delta x_i} & \frac{1}{\Delta x_i^2}
\end{pmatrix}.
\qquad (3.12)
$$

We may now solve (3.11) for the desired coefficients,

$$
A = [V(x_i)]^{-1}C[V(y_j)]^{-t},
\qquad (3.13)
$$

and so we can compute A by the multiplication of three 3×3 matrices. This is implemented in the Fortran subroutine QUM2D. Explicitly, the equations are

$$
\begin{aligned}
a_{11} &= u_{ij}, \\
a_{12} &= q_{ij}, \\
a_{13} &= \frac{1}{\Delta y_j}\left(\frac{u_{i,j+1}-u_{ij}}{\Delta y_j}-q_{ij}\right), \\
a_{21} &= p_{ij}, \\
a_{22} &= r_{ij}, \\
a_{23} &= \frac{1}{\Delta y_j}\left(\frac{p_{i,j+1}-p_{ij}}{\Delta y_j}-r_{ij}\right), \\
a_{31} &= \frac{1}{\Delta x_i}\left(\frac{u_{i+1,j}-u_{ij}}{\Delta x_i}-p_{ij}\right), \\
a_{32} &= \frac{1}{\Delta x_i}\left(\frac{q_{i+1,j}-q_{ij}}{\Delta x_i}-r_{ij}\right), \\
a_{33} &= \frac{1}{\Delta x_i \Delta y_j}\left[\frac{u_{i+1,j+1}-u_{i,j+1}-u_{i+1,j}+u_{ij}}{\Delta x_i \Delta y_j}-\frac{p_{i,j+1}-p_{ij}}{\Delta y_j}\right. \\
&\qquad\qquad \left. -\frac{q_{i+1,j}-q_{ij}}{\Delta x_i}+r_{ij}\right].
\end{aligned}
\tag{3.14}
$$

We now return to the problem of determining all the values p_{ij}, $i = 2,\cdots,n-1; j = 1,\cdots,m-1$, q_{ij}, $i = 1,\cdots,n-1; j = 2,\cdots,m$, and r_{ij}, $i = 1,\cdots,n-1; j = 1,\cdots,m-1$ except $i=j=1$. These together with (3.8) would allow all the required coefficients, $a_{ijk\ell}$, $i = 1,\cdots,n-1; j = 1,\cdots,m-1; k = 1,2,3; \ell = 1,2,3$, to be determined. The global function F is to have continuous first partial derivatives on all of R. We will guarantee this at first only for the grid points and subsequently show that it then automatically holds along the sides of neighboring rectangles. In the interior of R_{ij}, there is nothing to show. Since the form (3.3) is symmetric in x and y, we may restrict ourselves to showing the C^1 property for $y = y_j, j = 1,\cdots,m$, in the x direction and then conclude it by analogy for $x = x_i$, $i = 1,\cdots,n$, in the y direction.

By (3.3), for $y = y_j$,

$$
F_{ij}(x,y_j) = a_{ij11} + a_{ij21}(x-x_i) + a_{ij31}(x-x_i)^2
$$

and

$$
F_{i+1,j}(x,y_j) = a_{i+1,j,1,1} + a_{i+1,j,2,1}(x-x_{i+1}) + a_{i+1,j,3,1}(x-x_{i+1})^2.
$$

The C^1 requirement in the x direction,

$$
\frac{\partial F_{ij}}{\partial x}(x_{i+1},y_j) = \frac{\partial F_{i+1,j}}{\partial x}(x_{i+1},y_j),
$$

gives

$$a_{ij21} + 2a_{ij31}\Delta x_i = a_{i+1,j,2,1}.$$

Then substituting the correctly indexed coefficients of (3.14) yields the linear system of equations (with bidiagonal coefficient matrix),

$$p_{i+1,j} + p_{ij} = 2\frac{u_{i+1,j} - u_{ij}}{\Delta x_i}, \quad i = 1, \cdots, n - 1. \qquad (3.15)$$

For each $j = 1, \cdots, m$, this is easily solved for the desired p_{ij}, $i = 2, \cdots, n - 1$, by recursion down to the boundary values given by (3.8). Similarly, for $i = 1, \cdots, n$, we have the systems,

$$q_{i,j+1} + q_{ij} = 2\frac{u_{i,j+1} - u_{ij}}{\Delta y_j}, \quad j = 1, \cdots, m - 1, \qquad (3.16)$$

with given values q_{i1} to solve.

Now we still need formulas from which to be able to calculate the r_{ij}. According to (3.8), r_{11} was assumed to be given. Hence, it suffices to be able to compute the values $r_{i+1,j}$ and $r_{i,j+1}$ given r_{ij}. But by (3.6),

$$r_{i+1,j} = G_{xy}(x_{i+1}, y_j) = a_{22} + 2a_{32}\Delta x_i$$

and

$$r_{i,j+1} = G_{xy}(x_i, y_{j+1}) = a_{22} + 2a_{23}\Delta y_j.$$

The substitution of (3.14) yields the conditions,

$$r_{i+1,j} + r_{ij} = 2\frac{q_{i+1,j} - q_{ij}}{\Delta x_i}, \quad i = 1, \cdots, n - 1, \qquad (3.17)$$

and

$$r_{i,j+1} + r_{ij} = 2\frac{p_{i,j+1} - p_{ij}}{\Delta y_j}, \quad j = 1, \cdots, m - 1. \qquad (3.18)$$

Then either (3.17) for $j = 1$ and (3.18) for $i = 1, \cdots, n$, or (3.18) for $i = 1$ and (3.17) for $j = 1, \cdots, m$, can be used to compute all the required r_{ij}, $i = 1, \cdots, n - 1$; $j = 1, \cdots, m - 1$ except $i = j = 1$. At this point, all of the required coefficients $a_{ijk\ell}$, $i = 1, \cdots, n - 1$; $j = 1, \cdots, m - 1$; $k = 1, 2, 3$; $\ell = 1, 2, 3$, of the desired function F can be computed by means of (3.13) or (3.14). Furthermore, this shows that F exists and is unique. We need still to show that F really does have continuous first partials on the interior grid lines of the rectangular grid R. Again, by symmetry, we need only show that this is true in the x direction. Here, we will give an elementary argument. In later sections, we will give somewhat more elegant arguments for this type of problem. Hence, we consider two horizontally neighboring subrectangles R_{ij} and $R_{i+1,j}$ and investigate $\partial F / \partial x$ along the common side $x = x_{i+1}$ with $y_j < y < y_{j+1}$. By (3.4),

$$\frac{\partial F_{ij}}{\partial x}(x_{i+1}, y) = a_{ij21} + a_{ij22}(y - y_j) + a_{ij23}(y - y_j)^2$$
$$+ 2a_{ij31}\Delta x_i + 2a_{ij32}(y - y_j)\Delta x_i + 2a_{ij33}(y - y_j)^2 \Delta x_i$$

and

$$\frac{\partial F_{i+1,j}}{\partial x}(x_{i+1}, y) = a_{i+1,j,2,1} + a_{i+1,j,2,2}(y - y_j) + a_{i+1,j,2,3}(y - y_j)^2.$$

Substituting in the appropriately indexed values given by (3.14) on the right and then equating the left-hand sides yields the equation,

$$p_{ij} + r_{ij}(y - y_j) + \frac{1}{\Delta y_j}\left(\frac{p_{i,j+1} - p_{ij}}{\Delta y_j} - r_{ij}\right)(y - y_j)^2$$

$$+ 2\left(\frac{u_{i+1,j} - u_{ij}}{\Delta x_i} - p_{ij}\right) + 2\left(\frac{q_{i+1,j} - q_{ij}}{\Delta x_i} - r_{ij}\right)(y - y_j)$$

$$+ \frac{2}{\Delta y_j}\left(\frac{u_{i+1,j+1} - u_{i,j+1} - u_{i+1,j} + u_{ij}}{\Delta x_i \Delta y_j} - \frac{p_{i,j+1} - p_{ij}}{\Delta y_j}\right.$$
$$\left. - \frac{q_{i+1,j} - q_{ij}}{\Delta x_i} + r_{ij}\right)(y - y_j)^2 \tag{3.19}$$

$$= p_{i+1,j} + r_{i+1,j}(y - y_j) + \frac{1}{\Delta y_j}\left(\frac{p_{i+1,j+1} - p_{i+1,j}}{\Delta y_j} - r_{i+1,j}\right)(y - y_j)^2.$$

By equating the coefficients of 1, $(y - y_j)$, and $(y - y_j)^2$ on both sides, we see, by (3.15), (3.16), (3.17), and (3.18), that the asserted equality does indeed hold.

We now come to the implementation of the method. The boundary values (3.8) may either be approximated by simple difference quotients (ICASE=-1), explicitly given (ICASE=0), or determined in a certain sense optimally according to one of five different cases ([100]) (ICASE=$1, \cdots, 5$). These options are handled by the subroutine QUOPT2 (Fig. 3.1), which is simply a modified version of QUAOPT with the addition of the two cases ICASE=$-1, 0$. The subroutine QUM2D (Figs. 3.2 and 3.3) (ICASE here is called IR) solves the systems (3.16), (3.17), (3.18) for $j = 1$, and (3.19) for $i = 1, \cdots, n - 1$, with the help of QUOPT2. LAMBDA must be given for ICASE=IR=3 ([100]), and for ICASE=5, weights wx_i, $i = 1, \cdots, n$, in the x direction and wy_j, $j = 1, \cdots, m$ in the y direction, are required. For reasons of clarity and uniformity (see the subsequent sections), QUM2D then uses (3.13) to calculate the required coeffiecients $a_{ijk\ell}$. For this, the nonconstant elements of $[V(x_i)]^{-1}$ and $[V(y_j)]^{-1}$ are set by QUMMAT (Fig. 3.4).

To evaluate $G = F_{ij}$ of (3.3), we use a *bivariate version of Horner's rule*. For $(v, w) \in R_{ij}$ set $s = v - x_i$ and $t = w - y_j$. Then this method is

$$
\begin{aligned}
F_{ij}(v, w) &= a_{ij11} + t(a_{ij12} + ta_{ij13}) \\
&\quad + s[a_{ij21} + t(a_{ij22} + ta_{ij23}) + s(a_{ij31} + \\
&\quad t(a_{ij32} + ta_{ij33}))].
\end{aligned} \tag{3.20}
$$

In QUMVAL (Fig. 3.5), for $(v, w) \in R$ (error flag IFLAG=3 if $(v, w) \notin R$), INTTWO is first called to find the correct subrectangle and then (3.20) is evaluated. (Notation: QUMVAL= $F_{ij}(v, w)$, UX= s, UY= t.)

In the univariate case, we had the best results with ICASE=4 ([100]). Hence, this option was also used in Figs. 3.6–3.11. While the surface in Fig. 3.6 (with data taken from [64]) has an acceptable appearance, this is already less true of 3.7. The surfaces of Figs. 3.8–3.10 are inappropriate for smooth interpolation (other methods will produce much better results).

```
      SUBROUTINE QUOPT2(N,X,Y,ICASE,LAMBDA,W,A,B,C,IFLAG)
      DIMENSION X(N),Y(N),W(N),A(N),B(N),C(N)
      REAL LAMBDA
      IFLAG=0
      IF(N.LT.2) THEN
          IFLAG=1
          RETURN
      END IF
      IF(ICASE.LT.-1.OR.ICASE.GT.5) THEN
          IFLAG=5
          RETURN
      END IF
      N1=N-1
      DO 10 K=1,N1
          K1=K+1
          A(K)=X(K1)-X(K)
          C(K)=(Y(K1)-Y(K))/A(K)
 10   CONTINUE
      IF(ICASE.LE.0) GOTO 55
      IF(ICASE.EQ.1) LAMBDA=1.
      IF(ICASE.EQ.2) LAMBDA=0.
      IF(ICASE.LE.3) THEN
          P1=LAMBDA/3.
          P2=4.*(1.-LAMBDA)
          DO 20 K=1,N1
              AK=A(K)
              W(K)=P1*AK+P2/AK
 20       CONTINUE
      END IF
      IF(ICASE.EQ.4) THEN
          DO 30 K=1,N1
              CK=C(K)
              H=1.+CK*CK
              W(K)=4./(A(K)*H*H*H)
```

(cont.)

```
30      CONTINUE
        END IF
        WSUM=0.
        DO 40 K=1,N1
            WSUM=WSUM+W(K)
40      CONTINUE
        HSUM=WSUM
        ZSUM=0.
        VZ=1.
        DO 50 J=1,N1
            WJ=W(J)
            HSUM=HSUM-WJ
            ZSUM=ZSUM+VZ*(WJ+2.*HSUM)*C(J)
            VZ=-VZ
50      CONTINUE
        B(1)=ZSUM/WSUM
55      IF(ICASE.EQ.-1) B(1)=C(1)
        DO 60 K=1,N1
            K1=K+1
            BK=B(K)
            CK=C(K)
            B(K1)=2.*CK-BK
            C(K)=(CK-BK)/A(K)
            A(K)=Y(K)
60      CONTINUE
        RETURN
        END
```

Figure 3.1. Program listing of QUOPT2.

```
        SUBROUTINE QUM2D(N,M,NDIM,MDIM,MAXNM,X,Y,U,IR,LAMBDA,
       +                WX,WY,P,Q,R,A,IFLAG,UU,AA,BB,CC)
        DIMENSION X(N),Y(M),U(NDIM,M),WX(N),WY(M),P(NDIM,M),
       +          Q(NDIM,M),R(NDIM,M),A(NDIM,MDIM,3,3),
       +          UU(MAXNM),AA(MAXNM),BB(MAXNM),CC(MAXNM),
       +          B(3,3),C(3,3),D(3,3),E(3,3)
        REAL LAMBDA
        IFLAG=0
        IF(N.LT.2.OR.M.LT.2) THEN
            IFLAG=1
            RETURN
        END IF
        N1=N-1
        M1=M-1
        ZERO=0.
        DO 20 I=1,3
            DO 10 J=1,3
                B(I,J)=ZERO
                E(I,J)=ZERO
10          CONTINUE
20      CONTINUE
        B(1,1)=1.
        B(2,2)=1.
        E(1,1)=1.
```

(cont.)

```
      E(2,2)=1.
      DO 50 J=1,M
          IF(IR.EQ.0) BB(1)=P(1,J)
          DO 30 I=1,N
              UU(I)=U(I,J)
30        CONTINUE
          CALL QUOPT2(N,X,UU,IR,LAMBDA,WX,AA,BB,CC,IFLAG)
          IF(IFLAG.NE.0) RETURN
          DO 40 I=1,N1
              P(I,J)=BB(I)
40        CONTINUE
50    CONTINUE
      DO 80 I=1,N
          IF(IR.EQ.0) BB(1)=Q(I,1)
          DO 60 J=1,M
              UU(J)=U(I,J)
60        CONTINUE
          CALL QUOPT2(M,Y,UU,IR,LAMBDA,WY,AA,BB,CC,IFLAG)
          IF(IFLAG.NE.0) RETURN
          DO 70 J=1,M1
              Q(I,J)=BB(J)
70        CONTINUE
80    CONTINUE
      IF(IR.EQ.0) BB(1)=R(1,1)
      DO 90 I=1,N
          UU(I)=Q(I,1)
90    CONTINUE
      CALL QUOPT2(N,X,UU,IR,LAMBDA,WX,AA,BB,CC,IFLAG)
      IF(IFLAG.NE.0) RETURN
      DO 100 I=1,N1
          R(I,1)=BB(I)
100   CONTINUE
      DO 130 I=1,N1
          BB(1)=R(I,1)
          DO 110 J=1,M
              UU(J)=P(I,J)
110       CONTINUE
          CALL QUOPT2(M,Y,UU,0,LAMBDA,WY,AA,BB,CC,IFLAG)
          IF(IFLAG.NE.0) RETURN
          DO 120 J=1,M1
              R(I,J)=BB(J)
120       CONTINUE
130   CONTINUE
      DO 210 I=1,N1
          I1=I+1
          CALL QUMMAT(X(I),X(I1),B)
          DO 200 J=1,M1
              J1=J+1
              C(1,1)=U(I,J)
              C(1,2)=Q(I,J)
              C(1,3)=U(I,J1)
              C(2,1)=P(I,J)
              C(2,2)=R(I,J)
              C(2,3)=P(I,J1)
              C(3,1)=U(I1,J)
              C(3,2)=Q(I1,J)
              C(3,3)=U(I1,J1)
```

(cont.)

```
            DO 160 K1=1,3
               DO 150 K2=1,3
                  SUM=ZERO
                  DO 140 K=1,3
                     SUM=SUM+B(K1,K)*C(K,K2)
140                  CONTINUE
                  D(K1,K2)=SUM
150            CONTINUE
160         CONTINUE
            CALL QUMMAT(Y(J),Y(J1),E)
            DO 190 K1=1,3
               DO 180 K2=1,3
                  SUM=ZERO
                  DO 170 K1=1,3
                     SUM=SUM+D(K1,K)*E(K2,K)
170                  CONTINUE
                  A(I,J,K1,K2)=SUM
180            CONTINUE
190         CONTINUE
200      CONTINUE
210 CONTINUE
    RETURN
    END
```

Figure 3.2. Program listing of QUM2D.

Calling sequence:

CALL QUM2D(N,M,NDIM,MDIM,MAXNM,X,Y,U,IR,LAMBDA,
         WX,WY,P,Q,R,A,IFLAG,UU,AA,BB,CC)

Purpose:

For given points (x_i, y_j), $i = 1, \cdots, n \geq 2$; $j = 1, \cdots, m \geq 2$, in the plane with $x_1 < \cdots < x_n$ and $y_1 < \cdots < y_m$ and associated heights u_{ij}, this routine determines the coefficients $a_{ijk\ell}$ of a biquadratic spline interpolant F, i.e., on each subrectangle R_{ij}, F is of the form (1.12) with $N = 3$.

The boundary values needed for this calculation can be optionally computed by one of seven methods. These are selected through the variable IR. For IR=0, the boundary values for the first partial derivative with respect to x, p_{1j}, $j = 1, \cdots, m$, with respect to y, q_{i1}, $i = 1, \cdots, n$ and the value r_{11} of the mixed partial at the point (x_1, y_1), must be supplied by the user. For IR=-1,1,2,3,4,5, the boundary values are determined in a certain optimal manner (see the text) by means of QUOPT2. For IR=5, weights $wx_i > 0$, $i = 1, \cdots, n$, and $wy_j > 0$, $j = 1, \cdots, m$, must also be supplied. For IR=-1, the boundary values are determined by the formation of simple difference quotients. For IR=3, a $\lambda \in (0, 1)$ must be supplied.

Description of the parameters:

N,M,NDIM,X,Y,U as before.

MDIM	Maximum second dimnsion of A. Restriction: MDIM$\geq$M.
MAXNM	Maximum of M and N.
IR	=0: Boundary values are supplied by the user.
	=−1,1,2,3,4,5: Determination of the boundary values by QUOPT2.
LAMBDA	Must be supplied for IR=3. Restriction: 0<LAMBDA<1.
WX	ARRAY(N): Weights $wx_i > 0$. (Must be given for IR=5, cf. the text and QUOPT2.)
WY	ARRAY(M): Weights $wy_j > 0$. (Must be given for IR=5.)
P	ARRAY(NDIM,M): Output: Partial derivatives with respect to x.
Q	ARRAY(NDIM,M): Output: Partial derivatives with respect to y.
R	ARRAY(NDIM,M): Output: Mixed partial derivatives.
A	ARRAY(NDIM,MDIM,3,3): Output: Coefficients of the spline interpolant.
IFLAG	=0: Normal execution.
	=1: N< 2 or M<2.
	=5: IR< −1 or IR>5.

UU,AA,BB,CC ARRAY(MAXNM): Work space.

Required subroutines: QUOPT2, QUMMAT.

Figure 3.3. Description of QUM2D.

```
SUBROUTINE QUMMAT(H1,H2,B)
DIMENSION B(3,3)
H3=1./(H2-H1)
H4=H3*H3
B(3,1)=-H4
B(3,2)=-H3
B(3,3)=H4
RETURN
END
```

Figure 3.4. Program listing of QUMMAT.

```
FUNCTION QUMVAL(U,V,N,NDIM,M,MDIM,X,Y,A,IFLAG)
DIMENSION X(N),Y(M),A(NDIM,MDIM,3,3)
DATA I,J/2*1/
ZERO=0.
IFLAG=0
IF(N.LT.2.OR.M.LT.2) THEN
    IFLAG=1
    RETURN
END IF
CALL INTTWO(X,N,Y,M,U,V,I,J,IFLAG)
IF(IFLAG.NE.0) RETURN
UX=U-X(I)
VY=V-Y(J)
QUMVAL=ZERO
DO 20 K=3,1,-1
    QU=A(I,J,K,3)
    DO 10 L=2,1,-1
        QU=QU*VY+A(I,J,K,L)
10      CONTINUE
        QUMVAL=QUMVAL*UX+QU
20  CONTINUE
    RETURN
    END
```

FUNCTION QUMVAL(V,W,N,NDIM,M,MDIM,X,Y,A,IFLAG)

Purpose:
Calculation of the function value of a biquadratic spline interpolant
(cf. program description of QUM2D) at the point $(v, w) \in R$, where R
denotes the underlying rectangular grid.

Description of the parameters:

N,NDIM,M,MDIM,X,Y as in QUM2D.

V	x-coordinate of the point of evaluation.
W	y-coordinate of the point of evaluation.
A	ARRAY(NDIM,MDIM,3,3): Coefficients of the spline interpolant. These are to be computed outside of QUMVAL by the subroutine QUM2D.
IFLAG	=0: Normal execution.
	=1: N<2 or M<2.
	=3: Error in INTTWO.

Required subroutine: INTTWO.

Figure 3.5. Program listing of QUMVAL and its description.

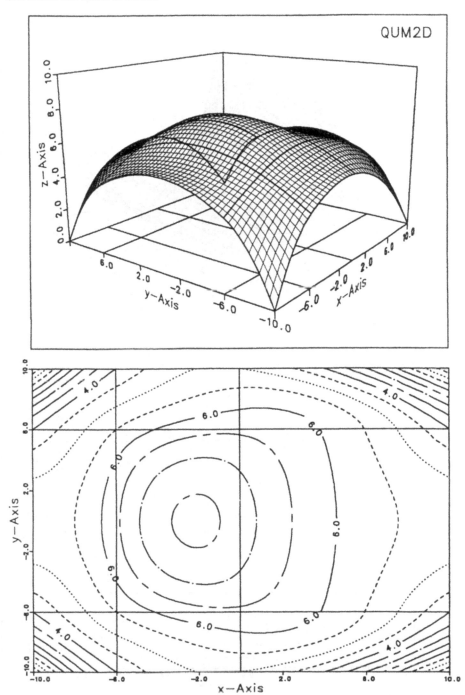

Figure 3.6.

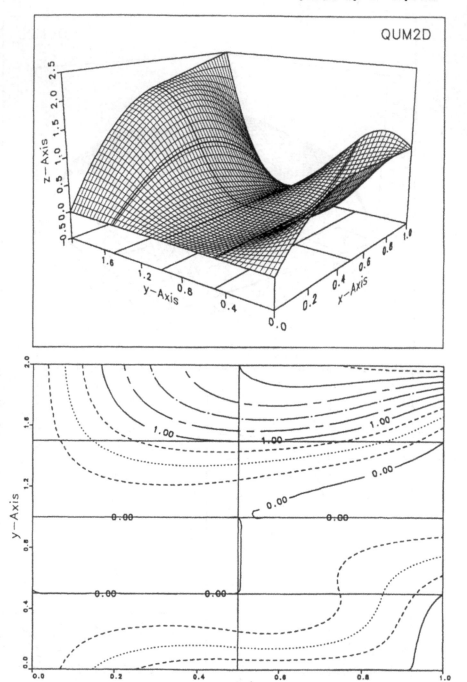

Figure 3.7.

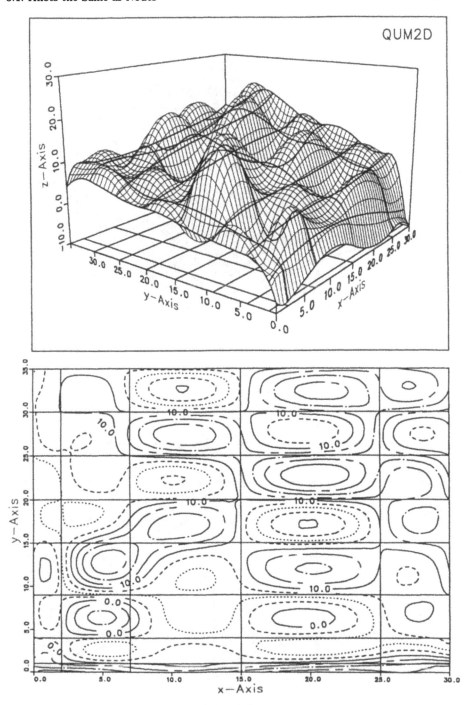

Figure 3.8.

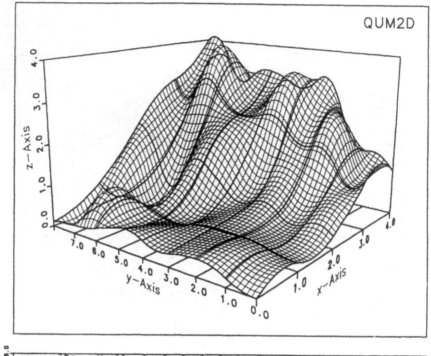

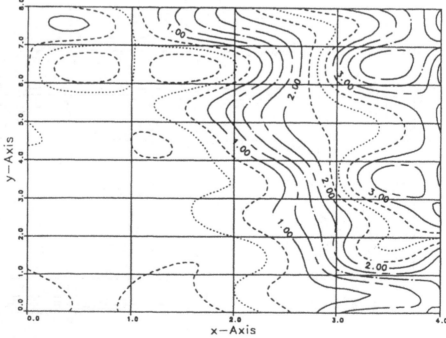

Figure 3.9.

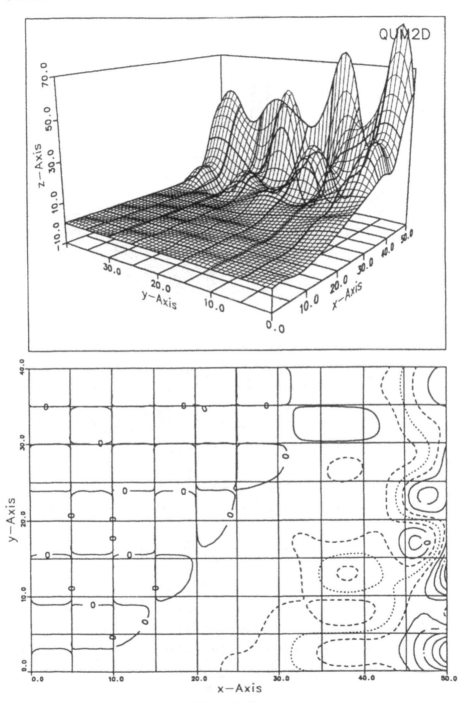

Figure 3.10.

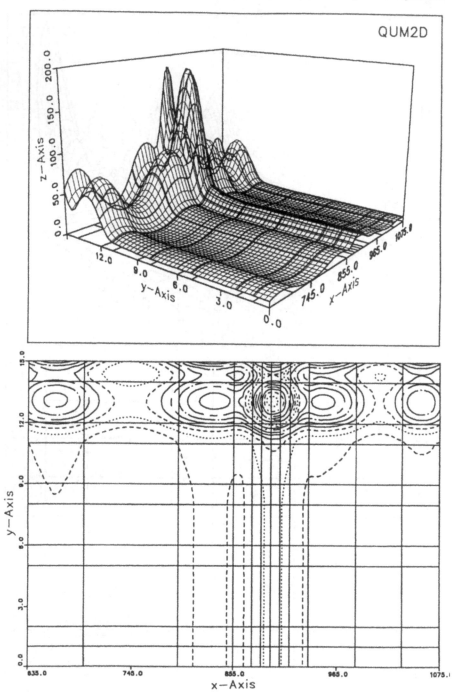

Figure 3.11.

Other boundary conditions or even a two-dimensional variant of ICASE=4 ([50]) are hardly any better. As in univariate interpolation, it appears that here also, C^1 biquadratic spline interpolants with knots the same as nodes are almost always unusable. Essentially, this is due to the one-sided, asymmetric boundary conditions. (Initial boundary value problems are also numerically more unstable than are boundary value problems also for differential equations.)

3.2. Knots Different from Nodes

In the univariate case, we had much better success with C^1 quadratic splines when the the knots were chosen to be intermediate to the nodes. We could even, by judicious choice of the knots, obtain shape-preserving properties ([100]). Often they were just as good as or even better than C^2 cubic splines. As we now want to apply *product interpolation* to this situation, the knots in two dimensions must lie on a rectangular grid R', which is again parallel to the axes. Hence, for $n > 2$ or $m > 2$, the knots on each subrectangle R'_{ij} cannot be chosen independently of those on the neighboring subrectangles. Knot coordinates $z_0, \cdots, z_n$ and $t_0, \cdots, t_m$ in the x and y directions, respectively, satisfying

$$z_0 < x_1 < z_1 < x_2 < z_2 < \cdots < z_{n-1} < x_n < z_n,$$

$$(3.21)$$

$$t_0 < y_1 < t_1 < y_2 < t_2 < \cdots < t_{m-1} < y_m < t_m,$$

are possible ([66]). For each i, z_i enters into all the vertical subrectangles R'_{ij}, and similarly t_j enters into all the horizontal subrectangles R'_{ij}. Since it is not possible to choose the z_i and t_j to simultaneously be in some sense optimal for each individual subrectangle, we restrict ourselves to the special case ([77]) of fixing

$$z_i = \frac{x_i + x_{i+1}}{2}, \quad i = 1, \cdots, n-1,$$

$$(3.22)$$

$$t_i = \frac{y_j + y_{j+1}}{2}, \quad j = 1, \cdots, m-1,$$

with, in addition,

$$z_0 = x_1 - \frac{1}{2}\Delta x_1, \ z_n = x_n + \frac{1}{2}\Delta x_{n-1},$$

$$(3.23)$$

$$t_0 = y_1 - \frac{1}{2}\Delta y_1, \ t_m = y_m + \frac{1}{2}\Delta y_{m-1}.$$

(In [89], the x_i and y_j are also assumed to be equally spaced.) The $z_0, \cdots, z_n$ and $t_0, \cdots, t_m$ then together only define subrectangles R'_{ij}, $i = 0, \cdots, n-1$; $j = 0, \cdots, m-1$, whose midpoints are the grid points of the original grid and whose own grid points are

$$(z_i, t_j),\ (z_{i+1}, t_j),\ (z_{i+1}, t_{j+1}),\ (z_i, t_{j+1}). \tag{3.24}$$

These latter will now be our knots. In contrast to the last section, there is now only one interpolation requirement per rectangle R'_{ij}, namely, at the midpoint (x_i, y_j). As unknowns, it is then convenient ([100]) to use the (now) partial derivatives in the x direction at the midpoints of the vertical knots, i.e., at (z_{i-1}, y_j) and (z_i, y_j), and in the y direction at the midpoints of the horizontal knots, i.e., at (x_i, t_{j-1}) and (x_i, t_j). We will denote the values of these derivatives by $p'_{i-1,j}$, p'_{ij}, $q'_{i,j-1}$, and q'_{ij}, respectively. (The prime on these and subsequent unknowns is to remind the reader that they are about points on the boundary of R'_{ij} and not of R_{ij}.) We now still require four temporary quantities to be used in determining the nine coefficients of the functions (3.1) or (3.3), i.e., $G' = F'_{ij}$, which are now defined relative to R'_{ij}. (Here, the prime is to indicate that the coefficients of (3.3) also have a prime.) From our experience in the previous section, these quantities should be mixed partial derivatives. The values $r'_{i-1,j-1}$, $r'_{i,j-1}$, r'_{ij}, and $r'_{i-1,j}$, $i = 1, \cdots, m$; $j = 1, \cdots, m$, corresponding to the corners of the R'_{ij}, are a convenient choice. With these, the matrix corresponding to (3.9), appropriate in the present circumstance, is

$$C' = \begin{pmatrix} r'_{i-1,j-1} & p'_{i-1,j} & r'_{i-1,j} \\ q'_{i,j-1} & u_{ij} & q'_{ij} \\ r'_{i,j-1} & p'_{ij} & r'_{ij} \end{pmatrix}. \tag{3.25}$$

For the all together $9nm$ coefficients, $a'_{ijk\ell}$, $i = 1, \cdots, n$; $j = 1, \cdots, m$; $k = 1, 2, 3$; $\ell = 1, 2, 3$, we have now temporarily introduced $3nm + 2(n+m) + 1$ unknowns, p'_{ij}, $i = 0, \cdots, n$; $j = 1, \cdots, m$, q'_{ij}, $i = 1, \cdots, n$; $j = 0, \cdots, m$, and r'_{ij}, $i = 0, \cdots, n$; $j = 0, \cdots, m$. We now want to find for each subrectangle R'_{ij}, a dependence formula corresponding to (3.11). Seeing that

$$z_{i-1} - x_i = -\frac{1}{2}\Delta x_{i-1},\ z_i - x_i = \frac{1}{2}\Delta x_i \quad (\Delta x_0 = \Delta x_1, \Delta x_n = \Delta x_{n-1}),$$

$$t_{j-1} - y_j = -\frac{1}{2}\Delta y_{j-1},\ t_j - y_j = \frac{1}{2}\Delta y_j \quad (\Delta y_0 = \Delta y_1, \Delta y_m = \Delta y_{m-1}),$$

we obtain the equation,

$$C' = V'(x_i)A'[V'(y_j)]^t, \tag{3.26}$$

analogous to (3.11). Here, the connection matrix is

$$
V'(x_i) = \begin{pmatrix} g'_{i1}(z_{i-1}) & g'_{i2}(z_{i-1}) & g'_{i3}(z_{i-1}) \\ g_{i1}(x_i) & g_{i2}(x_i) & g_{i3}(x_i) \\ g'_{i1}(z_i) & g'_{i2}(z_i) & g'_{i3}(z_i) \end{pmatrix} = \begin{pmatrix} 0 & 1 & -\Delta x_{i-1} \\ 1 & 0 & 0 \\ 0 & 1 & \Delta x_i \end{pmatrix},
$$
$$(3.27)$$

and we have denoted the desired coefficients of (3.3) by $A' = (a'_{k\ell})_{k,\ell=1,2,3}$. (The g_{ik} are defined in (3.2); g'_{ik} means the derivative with respect to x.)
   Since

$$
[V(x_i)]^{-1} = \begin{pmatrix} 0 & 1 & 0 \\ \dfrac{\Delta x_i}{\Delta x_{i-1}+\Delta x_i} & 0 & \dfrac{\Delta x_{i-1}}{\Delta x_{i-1}+\Delta x_i} \\ \dfrac{-1}{\Delta x_{i-1}+\Delta x_i} & 0 & \dfrac{1}{\Delta x_{i-1}+\Delta x_i} \end{pmatrix}
\qquad (3.28)
$$

exists, we immediately have

$$
A' = [V'(x_i)]^{-1} C' [V'(y_j)]^{-t}. \qquad (3.29)
$$

We now come to the problem of how to determine the values p'_{ij}, $i = 0, \cdots, n$; $j = 1, \cdots, m$, q'_{ij}, $i = 1, \cdots, n$; $j = 0, \cdots, m$, and r'_{ij}, $i = 0, \cdots, m$; $j = 0, \cdots, m$, required for the matrix A'_{ij}, so that the resulting global function F', defined piecewise by F'_{ij} on R'_{ij}, is once continuously differentiable on all of R'.
   Corresponding to the univariate case, we (at first) prescribe

$$
p'_{0j},\ p'_{nj},\quad j = 1, \cdots, m,
$$
$$(3.30)$$
$$
q'_{i0},\ q'_{im},\quad i = 1, \cdots, n,
$$

as boundary values. Equations to determine the remaining values then result from the C^1 requirement in the x direction for $j = 1, \cdots, m$,

$$
\frac{1}{\Delta x_{i-1} + \Delta x_i} p'_{i-1,j} + \frac{1}{\Delta x_i} \left[2 + \frac{\Delta x_{i-1}}{\Delta x_{i-1} + \Delta x_i} + \frac{\Delta x_{i+1}}{\Delta x_i + \Delta x_{i+1}} \right] p'_{ij}
$$
$$
+ \frac{1}{\Delta x_i + \Delta x_{i+1}} p'_{i+1,j} = \frac{4}{\Delta x_i} \frac{u_{i+1,j} - u_{ij}}{\Delta x_i}, \qquad (3.31)
$$
$$
i = 1, \cdots, n-1,
$$

and in the y direction for $i = 1, \cdots, n$,

$$
\frac{1}{\Delta y_{j-1} + \Delta y_j} q'_{i,j-1} + \frac{1}{\Delta y_j} \left[2 + \frac{\Delta y_{j-1}}{\Delta y_{j-1} + \Delta y_j} + \frac{\Delta y_{j+1}}{\Delta y_j + \Delta y_{j+1}} \right] q'_{ij}
$$
$$
+ \frac{1}{\Delta y_j + \Delta y_{j+1}} q'_{i,j+1} = \frac{4}{\Delta y_j} \frac{u_{i,j+1} - u_{ij}}{\Delta y_j}, \qquad (3.32)
$$
$$
j = 1, \cdots, m-1.
$$

We will not give any more of the details, as we did in the previous section immediately before (3.15), but refer to the derivation in the univariate case ([100]), which is equally valid for the bivariate case. Instead of the bidiagonal systems (3.15) and (3.16), we must now solve tridiagonal systems, which is about twice as expensive.

We want now to arrange for the continuity of $\partial^2 F'/\partial x \partial y$ at the interior grid points of R' supposing that the values,

$$r'_{00}, \; r'_{n0}, \; r'_{nm}, \; r'_{0m}, \tag{3.33}$$

at the four corners of R' are given. We consider the points of two horizontally neighboring subrectangles, R'_{ij} and $R'_{i+1,j}$, on the line $y = y_j$ and the grid line $y = t_j$ having x-coordinates, $z_{i-1}, x_i, z_i, z_{i+1}$, and z_{i+1}. The values,

$$y = y_j: \quad p'_{i-1,j}, \, u_{ij}, \, p'_{ij} \, u'_{i+1,j}, \, p'_{i+1,j},$$
$$y = t_j: \quad r'_{i-1,j}, \, q'_{ij}, \, r'_{ij}, \, q'_{i+1,j}, \, r'_{i+1,j},$$

are either given or have been (partly already computed) introduced as unknowns. The problem of continuous differentiability in the x direction of $\partial F'/\partial y$ at the point (z_i, t_j), i.e., that of having

$$\frac{\partial^2 F'_{ij}}{\partial y \partial x}(z_i, t_j) = \frac{\partial^2 F'_{i+1,j}}{\partial y \partial x}(z_i, t_j),$$

is the same as the problem of continuous differentiability in the x direction of F' at (z_i, y_j), i.e., of having

$$\frac{\partial F'_{ij}}{\partial x}(z_i, y_j) = \frac{\partial F'_{i+1,j}}{\partial x}(z_i, y_j)$$

when F' values are replaced by $\partial F'/\partial y$ values and $\partial F'/\partial x$ values are replaced by $\partial^2 F'/\partial y \partial x$ values. Thus, (3.31) holds with u replaced by q' and p' replaced by r'. Since we have only the boundary values (3.33) at our disposal, we can only use (3.31) for $j = 0$ and $j = m$. This yields the two systems of equations,

$$\frac{1}{\Delta x_{i-1} \Delta x_i} r'_{i-1,j} + \frac{1}{\Delta x_i} \left[2 + \frac{\Delta x_{i-1}}{\Delta x_{i-1} + \Delta x_i} + \frac{\Delta x_{i+1}}{\Delta x_i + \Delta x_{i+1}} \right] r'_{ij}$$
$$+ \frac{1}{\Delta x_i + \Delta x_{i+1}} r'_{i+1,j} = \frac{4}{\Delta x_i} \frac{q'_{i+1,j} - q'_{ij}}{\Delta x_i}, \tag{3.34}$$
$$i = 1, \cdots, n-1,$$

for the unknown values r'_{ij}, $i = 1, \cdots, n-1$; $j = 0, m$. Both systems have tridiagonal coefficient matrices. Once these values have been computed at the grid points of the upper and lower edges of R', we can apply for all the z_i, $i = 0, \cdots, n$, the same considerations in the y direction. By symmetry, we obtain for $i = 0, \cdots, n$, the systems of equations

$$
\frac{1}{\Delta y_{j-1}\Delta y_j}r'_{i-1,j} + \frac{1}{\Delta y_j}\left[2 + \frac{\Delta y_{j-1}}{\Delta y_{j-1}+\Delta y_j} + \frac{\Delta y_{j+1}}{\Delta y_j + \Delta y_{j+1}}\right]r'_{ij}
$$
$$
+\frac{1}{\Delta y_j + \Delta y_{j+1}}r'_{i+1,j} = \frac{4}{\Delta y_i}\frac{p'_{i+1,j} - p'_{ij}}{\Delta y_j}, \qquad (3.35)
$$
$$
j = 1, \cdots, m - 1,
$$

which give us the remaining unknowns r'_{ij}, $i = 0, \cdots, n$; $j = 1, \cdots, m - 1$. (Alternatively, we could have first solved (3.35) for $i = 0, n$, and then (3.34) for $j = 0, \cdots, m$.) We have now all the ingredients in order to be able to compute, by (3.29), all the coefficient matrices $A' = A'_{ij}$. Observe that there are really only two different coefficient matrices among the $m + 2n + 2$ linear systems (3.31), (3.32), and (3.34) that are to be solved. One of these is $(n - 1) \times (n - 1)$ and the other $(m - 1) \times (m - 1)$. Moreover, both are tridiagonal and strictly diagonally dominant ([100]). From this, the existence and uniqueness of F' follow.

Is the global function $F' = F'_{ij}|_{R'_{ij}}$, $i = 0, \cdots, n-1$; $j = 0, \cdots, m-1$, now continuously differentiable on all of R'? As the F'_{ij} agree on the common edge of any two adjacent subrectangles, $\partial F'/\partial x$ is continuous across horizontal grid lines and $\partial F'/\partial y$ across vertical ones. Further, by construction, the (by (3.6)) bilinear functions $\partial^2 F'_{ij}/\partial x \partial y$ form a bilinear spline, continuous on all of R'. On the sides of R'_{ij}, $\partial^2 F'/\partial x \partial y$ restricts to a linear in x along the horizontal sides and to a linear in y along the vertical ones. Integrating these with respect to x and y, respectively, we obtain quadratics that corrrespond to $\partial F'/\partial y$ along horizontal edges and to $\partial F'/\partial x$ along vertical edges, respectively. By the continuity of $\partial^2 F'/\partial x \partial y$, for two vertically adjacent subrectangles, the antiderivatives $\partial F'/\partial y$ can only differ in the constant across the common edge. Similarly, for two horizontally neighboring subrectangles, $\partial F'/\partial x$ can only differ in the constant across the common edge. But then, since on each interior horizontal edge they share the common value q'_{ij} and on each vertical edge the common value p'_{ij}, these constants must be the same. Hence, $\partial F'/\partial x$ and $\partial F'/\partial y$ are continuous across the edges and hence on all of R'.

The implementation QUA2D (Figs. 3.12 and 3.13) of C^1 biquadratic spline interpolation with knots at the midpoints of the nodes (and vice versa) closely resembles, in terms of its structure, the subroutine CUBTWO of [101] for C^2 bicubic spline interpolation. The two tridiagonal coefficient matrices of (3.31) and (3.34) (or, equivalently, of (3.32) and (3.35)) are assembled by QUADEC (Fig. 3.14). TRDISA (see the appendix) finds the LU decomposition of tridiagonal matrices. The various right-hand sides are set up by QUPERM (Fig. 3.15). Likewise, the solution of the various linear systems is carried out by TRDISB, which makes use of the previously computed LU decompositions. (TRDISA and TRDISB together have the same function as TRIDIS of [100] for a single right-hand side.) QUAMAT (Fig. 3.16) sets the non-constant elements of $[V'(x_i)]^{-1}$ or $[V'(y_j)]^{-1}$ of (3.28). The primes used to distinguish R' from R are suppressed in both the subroutines themselves and their program descriptions.

For IR=1, the boundary values (3.30) and (3.33) are automatically estimated in QUA2D by simple difference quotients (carried out by the loops DO 50, DO 60, and the four enclosed statements). For IR=2, the values preset in the corresponding vectors are used. The FUNCTION QBIVAL (Fig. 3.17) can be used for evaluation. For $(v, w) \in R'$, it again first calls INTTWO to determine the correct subrectangle R'_{ij} for which $(v, w) \in R'_{ij}$, and then, as before, the biquadratic is evaluated by a bivariate Horner's rule.

Some of the examples (IR=1!) of Figs. 3.18–3.25 are taken from datasets that we have already made use of, and others are from datasets that will arise later. The grid lines of R' (not R) within R are shown as well as their projections onto the interpolating surfaces. Some of these examples computed with QUA2D can be compared with those computed with QUM2D. The correspondences are: Fig. 3.6 with 3.21, 3.7 with 3.22, 3.8 with 3.23, 3.9 with 3.24, and finally, 3.11 with 3.25. In the first two cases, the results are very similar, but in the last three, the improvement of QUA2D from QUM2D becomes clear. Thus, QUA2D is always to be preferred, but we still need other interpolation methods, as can be seen, for example, from the excessive waviness of Fig. 3.25 in the y direction.

```
      SUBROUTINE QUA2D(N,M,NDIM,MDIM,X,Y,U,IR,EPS,Z,T,P,Q,R,A,
     +               IFLAG,DX,AX,BX,CX,RX,DY,AY,BY,CY,RY)
      DIMENSION X(N),Y(M),U(NDIM,M),Z(0:N),T(0:M),P(0:NDIM,0:M),
     +          Q(0:NDIM,0:M),R(0:NDIM,0:M),A(NDIM,MDIM,3,3),
     +          DX(0:N),AX(N),BX(N),CX(N),RX(N),
     +          DY(0:M),AY(M),BY(M),CY(M),RY(M),
     +          B(3,3),C(3,3),D(3,3),E(3,3)
      IFLAG=0
      IF(N.LT.2.OR.M.LT.2) THEN
          IFLAG=1
          RETURN
      END IF
      N1=N-1
      M1=M-1
      ZERO=0.
      DO 10 I=1,N1
          DX(I)=X(I+1)-X(I)
10    CONTINUE
      DX(0)=DX(1)
      DX(N)=DX(N1)
```

(cont.)

```
    DO 20 J=1,M1
        DY(J)=Y(J+1)-Y(J)
20  CONTINUE
    DY(0)=DY(1)
    DY(M)=DY(M1)
    DO 40 I=1,3
        DO 30 J=1,3
            B(I,J)=ZERO
            E(I,J)=ZERO
30      CONTINUE
40  CONTINUE
    B(1,2)=1.
    E(1,2)=1.
    IF(IR.EQ.2) GOTO 70
    DO 50 J=1,M
        P(0,J)=(U(2,J)-U(1,J))/DX(1)
        P(N,J)=(U(N,J)-U(N1,J))/DX(N1)
50  CONTINUE
    DO 60 I=1,N
        Q(I,0)=(U(I,2)-U(I,1))/DY(1)
        Q(I,M)=(U(I,M)-U(I,M1))/DY(M1)
60  CONTINUE
    R(0,0)=((P(0,2)-P(0,1))/DY(1)+(Q(2,0)-Q(1,0))/DX(1))/2.
    R(0,M)=((P(0,M)-P(0,M1))/DY(M1)+(Q(2,M)-Q(1,M))/DX(1))/2.
    R(N,0)=((P(N,2)-P(N,1))/DY(1)+(Q(N,0)-Q(N1,0))/DX(N1))/2.
    R(N,M)=((P(N,M)-P(N,M1))/DY(M1)+(Q(N,M)-Q(N1,M))/DX(N1))/2.
70  CALL QUADEC(N,X,Z,DX,AX,BX)
    CALL TRDISA(N1,AX,BX,CX,EPS,IFLAG)
    IF(IFLAG.NE.0) RETURN
    IVJ=0
    CALL QUPERM(IVJ,1,M,1,N,NDIM,N1,P,U,AX,BX,CX,DX,RX)
    CALL QUADEC(M,Y,T,DY,AY,BY)
    CALL TRDISA(M1,AY,BY,CY,EPS,IFLAG)
    IF(IFLAG.NE.0) RETURN
    IVJ=1
    CALL QUPERM(IVJ,1,N,1,M,NDIM,M1,Q,U,AY,BY,CY,DY,RY)
    IVJ=0
    CALL QUPERM(IVJ,0,M,M,N,NDIM,N1,R,Q,AX,BX,CX,DX,RX)
    IVJ=1
    CALL QUPERM(IVJ,0,N,1,M,NDIM,M1,R,P,AY,BY,CY,DY,RY)
    DO 150 I=1,N
        I1=I-1
        CALL QUAMAT(DX(I1),DX(I),B)
        DO 140 J=1,M
            J1=J-1
            C(1,1)=R(I1,J1)
            C(1,2)=P(I1,J)
            C(1,3)=R(I1,J)
            C(2,1)=Q(I,J1)
            C(2,2)=U(I,J)
            C(2,3)=Q(I,J)
            C(3,1)=R(I,J1)
            C(3,2)=P(I,J)
            C(3,3)=R(I,J)
            DO 100 K1=1,3
                DO 90 K2=1,3
                    SUM=ZERO
```

(cont.)

```
                        DO 80 K=1,3
                            SUM=SUM+B(K1,K)*C(K,K2)
80                      CONTINUE
                        D(K1,K2)=SUM
90                  CONTINUE
100             CONTINUE
                CALL QUAMAT(DY(J1),DY(J),E)
                DO 130 K1=1,3
                    DO 120 K2=1,3
                        SUM=ZERO
                        DO 110 K=1,3
                            SUM=SUM+D(K1,K)*E(K2,K)
110                     CONTINUE
                        A(I,J,K1,K2)=SUM
120                 CONTINUE
130             CONTINUE
140         CONTINUE
150 CONTINUE
    RETURN
    END
```

Figure 3.12. Program listing of QUA2D.

Calling sequence:

CALL QUA2D(N,M,NDIM,MDIM,X,Y,U,IR,EPS,Z,T,P,Q,R,A,
   IFLAG,DX,AX,BX,CX,RX,DY,AY,BY,CY,RY)

Purpose:

For given (x_i, y_j) $i = 1, \cdots, n \geq 2$; $j = 1, \cdots, m \geq 2$ in the plane with $x_1 < \cdots < x_n$ and $y_1 < \cdots < y_m$ and associated heights u_{ij}, this routine determines the coefficients $a_{ijk\ell}$ of a biquadratic spline interpolant with knots as in (3.22) and (3.23). For IR=1, the boundary values needed for this calculation are automatically computed by means of simple difference quotients. For IR=2, the boundary values for the partial derivatives with respect to x, p_{ij}, $i = 0, n$; $j = 1, \cdots, m$, and with respect to y, q_{ij}, $i = 1, \cdots, n$; $j = 0, m$, as well as the values r_{00}, r_{0m}, r_{n0}, and r_{nm} for the mixed partials, must be supplied by the user.

Description of the parameters:

N,M,NDIM,MDIM,X,Y,U as before.

IR   =1: Automatic computation of the boundary values by
            means of simple difference quotients.
            =2: The boundary values are supplied by the user.

EPS	See TRDISA (appendix).
Z	ARRAY(0:N): Output: Knots in the x direction.
T	ARRAY(0:M): Output: Knots in the y direction.
P	ARRAY(0:NDIM,0:M): Output: Partials with respect to x.
Q	ARRAY(0:NDIM,0:M): Output: Partials with respect to y.
R	ARRAY(0:NDIM,0:M): Output: Mixed partials.
A	ARRAY(NDIM,MDIM,3,3): Output: Coefficients of the biquadratic spline interpolant.
IFLAG	=0: Normal execution.
	=1: N<2 or M<2.
	=2: Error in solving the linear system (TRDISA).
DX	ARRAY(0:N): Work space.
DY	ARRAY(0:M): Work space.
AX,BX,CX,RX	ARRAY(N): Work space.
AY,BY,CY,RY	ARRAY(M): Work space.

Required subroutines: QUADEC, TRDISA, QUPERM, TRDISB, QUAMAT.

Remark: As QUPERM is called once with the array P as an argument and once with the array Q as an argument, for technical reasons these two arrays must have the same dimensions.

Figure 3.13. Description of QUA2D.

```
      SUBROUTINE QUADEC(N,X,Z,DX,AX,BX)
      DIMENSION X(N),Z(0:N),DX(0:N),AX(N),BX(N)
      N1=N-1
      H1=DX(0)
      H2=H1
      Z(0)=X(1)-H1/2.
      Z(N)=X(N)+DX(N)/2.
      DO 10 K=1,N1
          H3=DX(K+1)
          Z(K)=X(K)+H2/2.
          H4=1./(H2+H3)
          AX(K)=H4
          BX(K)=(2.+H1/(H1+H2)+H3*H4)/H2
          H1=H2
          H2=H3
10    CONTINUE
      RETURN
      END
```

Figure 3.14. Program listing of QUADEC.

```
      SUBROUTINE QUPERM(IVJ,JS,M,M1,N,NDIM,N1,P,U,AX,BX,CX,
     +                  DX,RX)
      DIMENSION P(0:NDIM,0:M),U(JS:NDIM,JS:M),
     +          AX(N),BX(N),CX(N),DX(0:N),RX(N)
      DO 40 J=JS,M,M1
         DO 20 I=1,N1
            IF(IVJ.NE.0) THEN
               RX(I)=4.*(U(J,I+1)-U(J,I))/(DX(I)*DX(I))
               IF(I.EQ.1) RX(I)=RX(I)-P(J,0)/(DX(0)+DX(1))
               IF(I.EQ.N1) RX(I)=RX(I)-P(J,N)/(DX(N1)+DX(N))
            ELSE
               RX(I)=4.*(U(I+1,J)-U(I,J))/(DX(I)*DX(I))
               IF (I.EQ.1) RX(I)=RX(I)-P(0,J)/(DX(0)+DX(1))
               IF (I.EQ.N1) RX(I)=RX(I)-P(N,J)/(DX(N1)+DX(N))
            END IF
 20      CONTINUE
         CALL TRDISB(N1,AX,BX,CX,RX)
         DO 30 I=1,N1
            IF(IVJ.EQ.0) P(I,J)=RX(I)
            IF(IVJ.NE.0) P(J,I)=RX(I)
 30      CONTINUE
 40   CONTINUE
      RETURN
      END
```

Figure 3.15. Program listing of QUPERM.

```
      SUBROUTINE QUAMAT(H1,H2,B)
      DIMENSION B(3,3)
      H3=1./(H1+H2)
      B(2,1)=H2*H3
      B(2,3)=H1*H3
      B(3,1)=-H3
      B(3,3)=H3
      RETURN
      END
```

Figure 3.16. Program listing of QUAMAT.

```
       FUNCTION QBIVAL(U,V,N,NDIM,M,MDIM,Z,T,X,Y,A,IFLAG)
       DIMENSION Z(N+1),T(M+1),X(N),Y(M),A(NDIM,MDIM,3,3)
       DATA I,J/2*1/
       ZERO=0.
       IFLAG=0
       IF(N.LT.2.OR.M.LT.2) THEN
           IFLAG=1
           RETURN
       END IF
       CALL INTTWO(Z,N+1,T,M+1,U,V,I,J,IFLAG)
       IF(IFLAG.NE.0) RETURN
       UX=U-X(I)
       VY=V-Y(J)
       QBIVAL=ZERO
       DO 20 K=3,1,-1
           QB=A(I,J,K,3)
           DO 10 L=2,1,-1
               QB=QB*VY+A(I,J,K,L)
10         CONTINUE
           QBIVAL=QBIVAL*UX+QB
20     CONTINUE
       RETURN
       END
```

+---+
| FUNCTION QBIVAL(V,W,N,NDIM,M,MDIM,Z,T,X,Y,A,IFLAG) |
| |
| Purpose: |
| Calculation of a function value of a biquadratic spline interpolant at |
| the point $(v, w) \in R'$, where R' denotes the rectangular grid defined in |
| (3.22) and (3.23) (knots at the midpoints of the nodes). |
| |
| Description of the parameters: |
| |
| V,W,N,NDIM,M,MDIM,X,Y,IFLAG as in QUMVAL. |
| Z   ARRAY(N+1): Input: x-coordinates of the knots, z_i. |
| (z_i is stored in Z(I+1) $I, i = 0, \cdots, n$.) |
| T   ARRAY(M+1): Input: y-coordinates of the knots, t_j. |
| (z_j is stored in Z(J+1) $J, j = 0, \cdots, m$.) |
| A   ARRAY(NDIM,MDIM,3,3): Coefficients of the spline |
| interpolant. |
| |
| Required subroutine: INTTWO. |
+---+

Figure 3.17. Program listing of QBIVAL and its description.

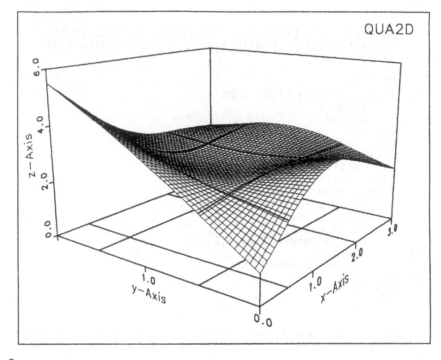

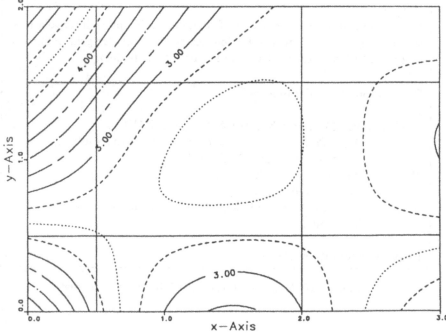

Figure 3.18.

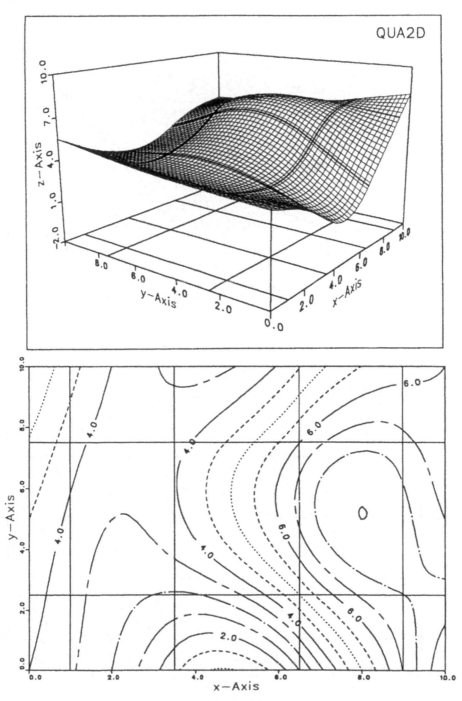

Figure 3.19.

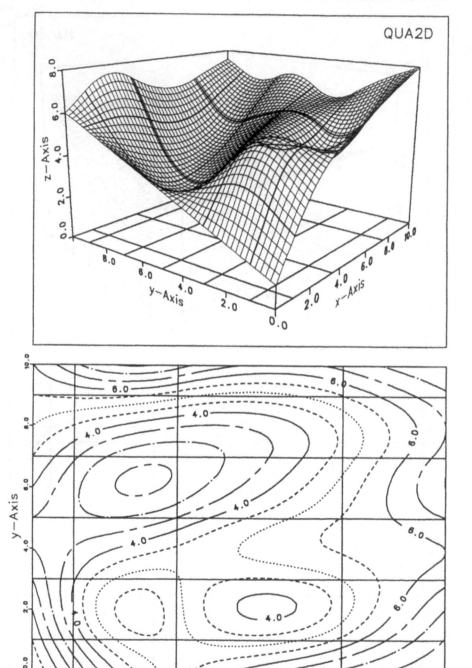

Figure 3.20.

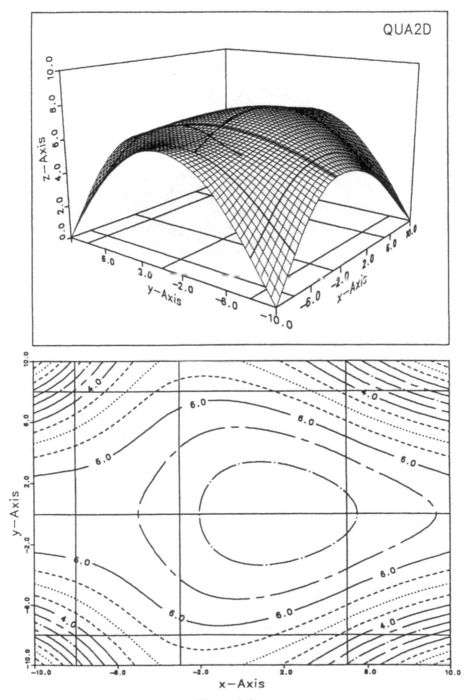

Figure 3.21.

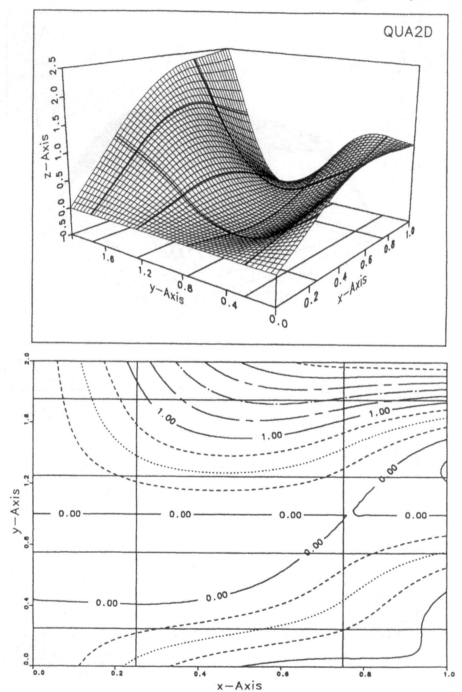

Figure 3.22.

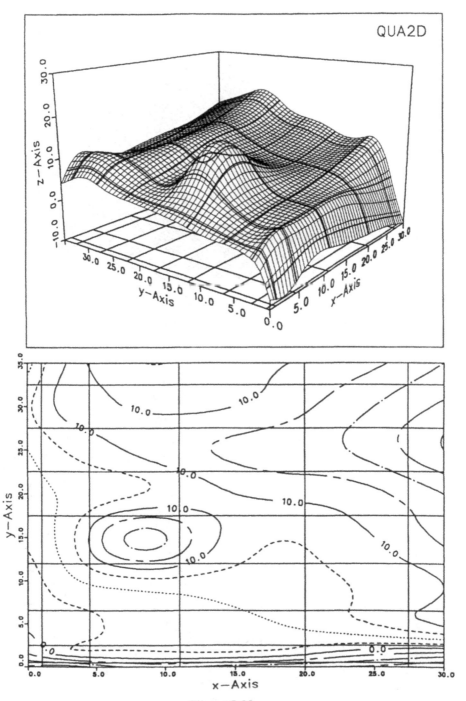

Figure 3.23.

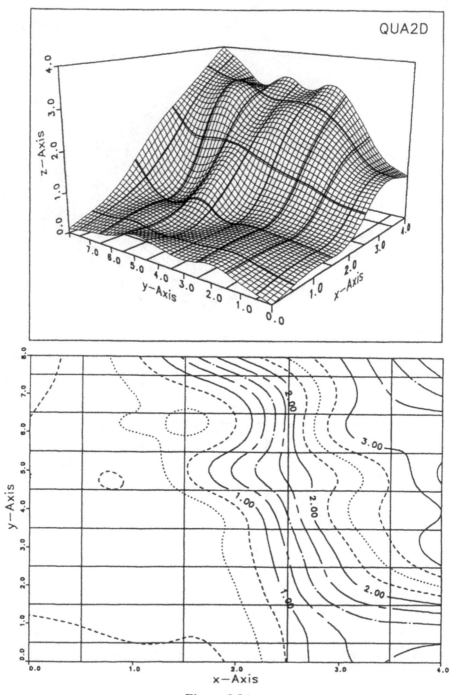

Figure 3.24.

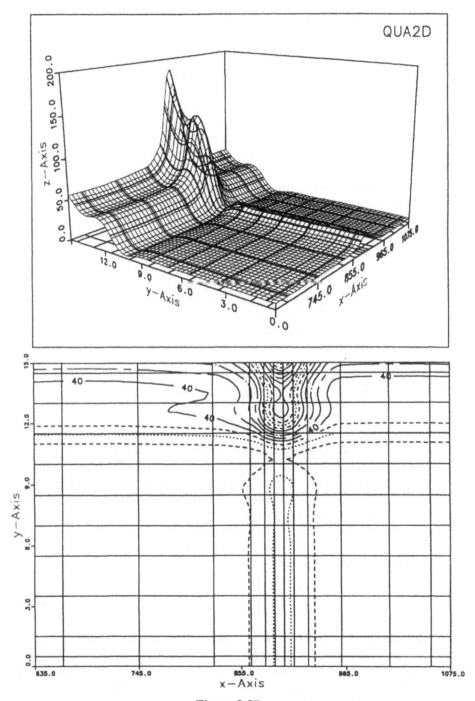

Figure 3.25.

3.3. Shape Preservation

Although, next to bilinears and quadratics, biquadratic spline interpolation on rectangular grids R or R' uses polynomials as low degree as possible, very few results on shape preservation, i.e., the reproduction of positivity, monotonicity, and convexity, seem to be known.

For monotone data, the subrectangles can be subdivided into 12 regular triangles and the first partials at the corners of the rectangle so determined that there results on each rectangle, and hence globally, a monotone C^1 biquadratic spline ([9]). In [77], each rectangle is subdivided into four triangles, on each of which is set a quadratic of the form (1.13) for $n = 2$. This allows contour lines to be easily computed. A certain kind of convexity for C^1 biquadratic splines is characterized in [91] by nonlinear equations in the p, q, and r. There seems to be little chance that this can be implemented algorithmically with any efficiency.

All of these, and other such methods, seem, in comparison to C^2 birational splines (with two prescribable real poles in the terminology of [100]) and suitable choice of poles, mathematically interesting but too complicated to be implemented. We will return later to the topic of shape preservation and then give some examples.

3.4. A Local Quadratic Method of Interpolation

The biquadratic spline interpolation methods just discussed yield surfaces that have continuous first partial derivatives in both the x and y directions. They are *global* methods, as evaluation at a given point in the rectangle, other than a grid point, always requires using *all* of the data. We wish to describe briefly a completely simple *local* quadratic method that, however, almost always does not even produce a continuous interpolant. A Fortran subroutine, QD2VL, for this method is available in the IMSL library.

We select a quadratic function,

$$G(x,y) = b_1 + b_2(x - x_i) + b_3(y - y_j)$$
$$+b_4(x - x_i)^2 + b_5(y - y_j)^2 + b_6(x - x_i)(y - y_j), \qquad (3.36)$$

of the type (1.13). The pair of indices (i, j) is chosen so that (x_i, y_j) is an *interior* grid point lying closest to the interpolation point (v, w). As additional nodes, we take the four grid points immediately neighboring (x_i, y_j) and then finally the closest of the four grid points $(x_{i\pm1}, y_{j\pm1})$ to (v, w). This last point we will denote by (x_k, y_ℓ). The requirement of interpolation at these six points then gives the following conditions for the six unknown coefficients:

$$
\begin{aligned}
u_{ij} &= b_1, \\
u_{i+1,j} &= b_1 + b_2 \Delta x_i + b_4 \Delta x_i^2, \\
u_{i-1,j} &= b_1 - b_2 \Delta x_{i-1} + b_4 \Delta x_{i-1}^2, \\
u_{i,j+1} &= b_1 + b_3 \Delta y_j + b_5 \Delta y_j^2, \\
u_{i,j-1} &= b_1 - b_3 \Delta y_{j-1} + b_5 \Delta y_{j-1}^2, \\
u_{k,\ell} &= b_1 + b_2 (x_k - x_i) + b_3 (y_\ell - y_j) + b_4 (x_k - x_i)^2 \\
&\quad + b_5 (y_\ell - y_j)^2 + b_6 (x_k - x_i)(y_\ell - y_j).
\end{aligned}
$$

From the first five equations, it immediately follows that

$$
\begin{aligned}
b_1 &= u_{ij}, \\
b_2 &= \frac{1}{\Delta x_{i-1} + \Delta x_i} \left[\Delta x_{i-1} \frac{u_{i+1,j} - u_{ij}}{\Delta x_i} + \Delta x_i \frac{u_{ij} - u_{i-1,j}}{\Delta x_{i-1}} \right], \\
b_4 &= \frac{1}{\Delta x_{i-1} + \Delta x_i} \left[\frac{u_{i+1,j} - u_{ij}}{\Delta x_i} - \frac{u_{ij} - u_{i-1,j}}{\Delta x_{i-1}} \right], \\
b_3 &= \frac{1}{\Delta y_{j-1} + \Delta y_j} \left[\Delta y_{j-1} \frac{u_{i,j+1} - u_{ij}}{\Delta y_j} + \Delta y_j \frac{u_{ij} - u_{i,j-1}}{\Delta y_{j-1}} \right], \\
b_5 &= \frac{1}{\Delta y_{j-1} + \Delta y_j} \left[\frac{u_{i,j+1} - u_{ij}}{\Delta y_j} - \frac{u_{ij} - u_{i,j-1}}{\Delta y_{j-1}} \right],
\end{aligned}
$$

and then b_6 can always be computed from the sixth equation. The global function F, formed in this way, can be used, for example, to estimate values for the first and possibly the second derivatives at points in the interior of the subrectangle R_{ij} that are not points of discontinuity.

It is also possible to construct C^1 quadratic Hermite splines of the type (3.36). However, this would involve splitting each subrectangle into six or 12 triangles.

4

Bicubic Spline Interpolation

4.1. Bicubic Spline Interpolation on Rectangular Grids

We now return again to the original rectangle R with grid points (1.2) as interpolation nodes. A bicubic function on a subrectangle R_{ij} has the form,

$$G(x,y) = (g_{i1}(x), g_{i2}(x), g_{i3}(x), g_{i4}(x))A \begin{pmatrix} g_{j1}(y) \\ g_{j2}(y) \\ g_{j3}(y) \\ g_{j4}(y) \end{pmatrix}, \qquad (4.1)$$

where the g_{ik} are those of (3.2) with the addition of

$$g_{i4}(x) = (x - x_i)^3, \qquad (4.2)$$

and the matrix,

$$A = \begin{pmatrix} a_{ij11} & a_{ij12} & a_{ij13} & a_{ij14} \\ a_{ij21} & a_{ij22} & a_{ij23} & a_{ij24} \\ a_{ij31} & a_{ij32} & a_{ij33} & a_{ij34} \\ a_{ij41} & a_{ij42} & a_{ij43} & a_{ij44} \end{pmatrix}, \qquad (4.3)$$

is now 4×4. We wish to determine the total of $16(n-1)(m-1)$ coefficients so that the global function, $F|_{R_{ij}} = F_{ij}$, formed by taking $F_{ij} = G$ is twice

71

continuously differentiable on all of R.

Since the interpolation requirements give four conditions, we still need to introduce a total of 12 temporary unknowns or parameters. A natural choice for these is that of the values of the first partials as well as the mixed partial derivative at each of the four nodes. In this case, the matrix (4.3) is to be determined from the matrix ([21]),

$$C = \begin{pmatrix} u_{ij} & q_{ij} & u_{i,j+1} & q_{i,j+1} \\ p_{ij} & r_{ij} & p_{i,j+1} & r_{i,j+1} \\ u_{i+1,j} & q_{i+1,j} & u_{i+1,j+1} & q_{i+1,j+1} \\ p_{i+1,j} & r_{i+1,j} & p_{i+1,j+1} & r_{i+1,j+1} \end{pmatrix}. \tag{4.4}$$

If we write the partial derivatives of (4.1) just as we did in (3.4) through (3.6), then we arrive at the *connection matrix*,

$$\begin{aligned} V(x_i) &= \begin{pmatrix} g_{i1}(x_i) & g_{i2}(x_i) & g_{i3}(x_i) & g_{i4}(x_i) \\ g'_{i1}(x_i) & g'_{i2}(x_i) & g'_{i3}(x_i) & g'_{i4}(x_i) \\ g_{i1}(x_{i+1}) & g_{i2}(x_{i+1}) & g_{i3}(x_{i+1}) & g_{i4}(x_{i+1}) \\ g'_{i1}(x_{i+1}) & g'_{i2}(x_{i+1}) & g'_{i3}(x_{i+1}) & g'_{i4}(x_{i+1}) \end{pmatrix} \\ &= \begin{pmatrix} 1 & 0 & 0 & 0 \\ 0 & 1 & 0 & 0 \\ 1 & \Delta x_i & \Delta x_i^2 & \Delta x_i^3 \\ 0 & 1 & 2\Delta x_i & 3\Delta x_i^2 \end{pmatrix}, \end{aligned} \tag{4.5}$$

i.e., (4.3) and (4.4) satisfy the relation,

$$C = V(x_i) A [V(y_j)]^t. \tag{4.6}$$

Since the inverse of $V(x_i)$ may explicitly be computed to be

$$[V(x_i)]^{-1} = \begin{pmatrix} 1 & 0 & 0 & 0 \\ 0 & 1 & 0 & 0 \\ -\frac{3}{\Delta x_i^2} & -\frac{2}{\Delta x_i} & \frac{3}{\Delta x_i^2} & -\frac{1}{\Delta x_i} \\ \frac{2}{\Delta x_i^3} & \frac{1}{\Delta x_i^2} & -\frac{2}{\Delta x_i^3} & \frac{1}{\Delta x_i} \end{pmatrix}, \tag{4.7}$$

we may solve (4.6) for

$$A = [V(x_i)]^{-1} C [V(y_j)]^{-t}. \tag{4.8}$$

Using the equations of univariate cubic spline interpolation ([100, formula (4.7)]), the C^2 requirements in the x and y directions yield, for $j = 1, \cdots, m$, the equations,

$$\frac{1}{\Delta x_{i-1}} p_{i-1,j} + 2 \left[\frac{1}{\Delta x_{i-1}} + \frac{1}{\Delta x_i} \right] p_{ij} + \frac{1}{\Delta x_i} p_{i+1,j}$$

$$= 3 \left[\frac{u_{ij} - u_{i-1,j}}{\Delta x_{i-1}^2} + \frac{u_{i+1,j} - u_{ij}}{\Delta x_i^2} \right], \quad i = 2, \cdots, n-1, \tag{4.9}$$

and for $i = 1, \cdots, n$, the equations,

$$\frac{1}{\Delta y_{j-1}} q_{i,j-1} + 2\left[\frac{1}{\Delta y_{j-1}} + \frac{1}{\Delta y_j}\right] q_{ij} + \frac{1}{\Delta y_j} q_{i,j+1}$$

$$= 3\left[\frac{u_{ij} - u_{i,j-1}}{\Delta y_{j-1}^2} + \frac{u_{i,j+1} - u_{ij}}{\Delta y_j^2}\right], \quad j = 2, \cdots, m-1. \tag{4.10}$$

These are analogous to (3.31) and (3.32). For these, we assume that we are given the values

$$p_{ij}, \quad i = 1, n; \quad j = 1, \cdots, m,$$

$$q_{ij}, \quad i = 1, \cdots, n; \quad j = 1, m, \tag{4.11}$$

of the first partial with respect to x at the grid points of both vertical edges of R and of the first partials with respect to y along both horizontal edges of R.

Further, we must arrange for the continuity of $\partial^4 F / \partial x^2 \partial y^2$ at all the interior grid points. If in addition to (4.11), we are given the values,

$$r_{11}, r_{n1}, r_{nm}, r_{1m}, \tag{4.12}$$

for the mixed partials at the corners of R, then, by the same reasoning as that following (3.33), we arrive at the systems,

$$\frac{1}{\Delta x_{i-1}} r_{i-1,j} + 2\left[\frac{1}{\Delta x_{i-1}} + \frac{1}{\Delta x_i}\right] r_{ij} + \frac{1}{\Delta x_i} r_{i+1,j}$$

$$= 3\left[\frac{q_{ij} - q_{i-1,j}}{\Delta x_{i-1}^2} + \frac{q_{i+1,j} - q_{ij}}{\Delta x_i^2}\right], \quad i = 2, \cdots, n-1, \tag{4.13}$$

for $j = 1$ and $j = m$, and then for $i = 1, \cdots, n$,

$$\frac{1}{\Delta y_{j-1}} r_{i,j-1} + 2\left[\frac{1}{\Delta y_{j-1}} + \frac{1}{\Delta y_j}\right] r_{ij} + \frac{1}{\Delta y_j} r_{i,j+1}$$

$$= 3\left[\frac{p_{ij} - p_{i,j-1}}{\Delta y_{j-1}^2} + \frac{p_{i,j+1} - p_{ij}}{\Delta y_j^2}\right], \quad j = 2, \cdots, m-1. \tag{4.14}$$

These $m + 2n + 2$ systems of equations may already be found in [21] and again have altogether only two different symmetric, diagonally dominant, tridiagonal coefficient matrices. Because these systems are uniquely solvable, the bicubic spline interpolant exists and is uniquely determined, just as was the biquadratic interpolant ([21]).

We need now to show that F is indeed C^2 across the interior edges of R. By construction $\partial^4 F / \partial x^2 \partial y^2$ is a bilinear spline interpolant, continuous on all of R. By integrating the corresponding linear functions along the sides of neighboring rectangles twice with respect to x, once each with respect to x and y, and twice with respect to y respectively, we obtain in each case polynomials on that edge whose two integration constants are the same because of the agreement of p-, r-, and q-values, respectively, at the edge endpoints.

The subroutine CUB2D (Figs. 4.1 and 4.2) corresponds to the formulas developed above. Its structure is based on [98,101]. The boundary values (4.11) and (4.12) may alternatively be preassigned (IR=2) or approximated by simple difference quotients (IR=1), just as in QUA2D. TRDISA is used to compute the LU decomposition of the tridiagonal coefficient matrix that arises in (4.9) and (4.13) as well as that of the one that arises in (4.10) and (4.14). CBPERM (Fig. 4.3) prepares the various right-hand sides and solves the corresponding systems by means of TRDISB and the output of TRDISA. CUBMAT (Fig. 4.4) generates the nonconstant elements of the 4×4 matrices, $[V(x_i)]^{-1}$ and $[V(y_j)]^{-1}$, for each i and j. Evaluation may be carried out by CBIVAL (Fig. 4.5), which works just like QBIVAL, except that the *bivariate Horner's rule* goes one index further.

In Figs. 4.6–4.17, IR is always IR=1, i.e., the boundary values are automatically set to the approximating difference quotients. Figures 4.6–4.11, 4.14, and 4.17 correspond to those for QUA2D given in Figs. 3.18–3.25. There is very little difference between these two sets of plots. The new examples 4.12, 4.13, 4.15([12]), 4.16([1,2]), and 4.17([19]) illustrate that bicubic spline interpolation (just as biquadratic) is occasionally somewhat deficient. Figures 4.15 and 4.17 are examples of monotone data. For these, it would be more appropriate to use one of the special methods developed for this purpose ([11,12,13,18,19,20,23,44]) or, as we prefer, to use birational spline interpolation. We will discuss this use in a later chapter. Summarizing, we can say that C^1 biquadratic spline interpolation (with knots between the nodes) is almost always just as good or bad as C^2 bicubic spline interpolation.

The choice of boundary conditions and corresponding values has, for larger m and n, no influence on the validity of this statement in the interior of R. Nevertheless, one could still, in principle, allow all of the possibilities that were implemented in CUB1R5 and CUB2R7([100]). These could even be different on each grid line or each side. Moreover, one could compute *natural bicubic splines* ([93]) or require *periodicity* on the boundary of the rectangle. Other boundary conditions involving two-dimensional *B-splines* are discussed in [96]. It has also been suggested that one omit some of our boundary values (4.11) and (4.12) and determine the remaining free

```
      SUBROUTINE CUB2D(N,M,NDIM,MDIM,X,Y,U,IR,EPS,P,Q,R,A,
     +              IFLAG,DX,AX,BX,CX,RX,DY,AY,BY,CY,RY)
      DIMENSION X(N),Y(M),U(NDIM,M),P(NDIM,M),Q(NDIM,M),
     +          R(NDIM,M),A(NDIM,MDIM,4,4),
     +          DX(N),AX(N),BX(N),CX(N),RX(N),
     +          DY(M),AY(M),BY(M),CY(M),RY(M),
     +          B(4,4),C(4,4),D(4,4),E(4,4)
      IFLAG=0
      IF(N.LT.2.OR.M.LT.2) THEN
          IFLAG=1
          RETURN
      END IF
      N1=N-1
      N2=N-2
      M1=M-1
      M2=M-2
      ZERO=0.
      DO 10 I=1,N1
          DX(I)=1./(X(I+1)-X(I))
10    CONTINUE
      DO 20 J=1,M1
          DY(J)=1./(Y(J+1)-Y(J))
20    CONTINUE
      DO 40 I=1,4
          DO 30 J=1,4
              B(I,J)=ZERO
              E(I,J)=ZERO
30        CONTINUE
40    CONTINUE
      B(1,1)=1.
      B(2,2)=1.
      E(1,1)=1.
      E(2,2)=1.
      IF(IR.EQ.2) GOTO 65
      DO 50 J=1,M
          P(1,J)=(U(2,J)-U(1,J))*DX(1)
          P(N,J)=(U(N,J)-U(N1,J))*DX(N1)
50    CONTINUE
      DO 60 I=1,N
          Q(I,1)=(U(I,2)-U(I,1))*DY(1)
          Q(I,M)=(U(I,M)-U(I,M1))*DY(M1)
60    CONTINUE
      R(1,1)=((P(1,2)-P(1,1))*DY(1)+(Q(2,1)-Q(1,1))*DX(1))/2.
      R(1,M)=((P(1,M)-P(1,M1))*DY(M1)+(Q(2,M)-Q(1,M))*DX(1))/2.
      R(N,1)=((P(N,2)-P(N,1))*DY(1)+(Q(N,1)-Q(N1,1))*DX(N1))/2.
      R(N,M)=((P(N,M)-P(N,M1))*DY(M1)+(Q(N,M)-Q(N1,M))*
     +      DX(N1))/2.
      IF(N.EQ.2.AND.M.EQ.2) GOTO 120
      IF(N.EQ.2) GOTO 80
65    DO 70 I=1,N2
          IF(I.LT.N2) AX(I)=DX(I+1)
          BX(I)=2.*(DX(I+1)+DX(I))
70    CONTINUE
      CALL TRDISA(N2,AX,BX,CX,EPS,IFLAG)
      IF(IFLAG.NE.0) RETURN
      IVJ=0
      CALL CBPERM(IVJ,M,1,N,NDIM,N1,N2,P,U,AX,BX,CX,DX,RX)
80    IF(M.EQ.2) GOTO 100
```

```
(cont.)
    DO 90 J=1,M2
        IF(J.LT.M2) AY(J)=DY(J+1)
        BY(J)=2.*(DY(J+1)+DY(J))
 90 CONTINUE
    CALL TRDISA(M2,AY,BY,CY,EPS,IFLAG)
    IF(IFLAG.NE.0) RETURN
    IVJ=1
    CALL CBPERM(IVJ,N,1,M,NDIM,M1,M2,Q,U,AY,BY,CY,DY,RY)
100 IF(N.EQ.2) GOTO 110
    IVJ=0
    CALL CBPERM(IVJ,M,M1,N,NDIM,N1,N2,R,Q,AX,BX,CX,DX,RX)
110 IF(M.EQ.2) GOTO 120
    IVJ=1
    CALL CBPERM(IVJ,N,1,M,NDIM,M1,M2,R,P,AY,BY,CY,DY,RY)
120 DO 200 I=1,N1
        I1=I+1
        CALL CUBMAT(DX(I),B)
        DO 190 J=1,M1
            J1=J+1
            C(1,1)=U(I,J)
            C(1,2)=Q(I,J)
            C(2,1)=P(I,J)
            C(2,2)=R(I,J)
            C(1,3)=U(I,J1)
            C(1,4)=Q(I,J1)
            C(2,3)=P(I,J1)
            C(2,4)=R(I,J1)
            C(3,1)=U(I1,J)
            C(3,2)=Q(I1,J)
            C(4,1)=P(I1,J)
            C(4,2)=R(I1,J)
            C(3,3)=U(I1,J1)
            C(3,4)=Q(I1,J1)
            C(4,3)=P(I1,J1)
            C(4,4)=R(I1,J1)
            DO 150 K1=1,4
                DO 140 K2=1,4
                    SUM=ZERO
                    DO 130 K=1,4
                        SUM=SUM+B(K1,K)*C(K,K2)
130                 CONTINUE
                    D(K1,K2)=SUM
140             CONTINUE
150         CONTINUE
            CALL CUBMAT(DY(J),E)
            DO 180 K1=1,4
                DO 170 K2=1,4
                    SUM=ZERO
                    DO 160 K=1,4
                        SUM=SUM+D(K1,K)*E(K2,K)
160                 CONTINUE
                    A(I,J,K1,K2)=SUM
170             CONTINUE
180         CONTINUE
190     CONTINUE
200 CONTINUE
    RETURN
    END
```

Figure 4.1. Program listing of CUB2D.

Calling sequence:

CALL CUB2D(N,M,NDIM,MDIM,X,Y,U,IR,EPS,P,Q,R,A,
            IFLAG,DX,AX,BX,CX,RX,DY,AY,BY,CY,RY)

Purpose:
For given points (x_i, y_j) $i = 1, \cdots, n \geq 2$; $j = 1, \cdots, m \geq 2$ in the plane
with $x_1 < \cdots < x_n$ and $y_1 < \cdots < y_m$ and associated heights u_{ij}, this
routine determines the coefficients $a_{ijk\ell}$ of a bicubic spline interpolant,
F, i.e., F is of the form (1.12) with $N = 4$ on each of the subrectangles
R_{ij}. For IR=1, the boundary values needed for this calculation are automatically computed by means of simple difference quotients. For IR=2,
the boundary values for the partial derivatives with respect to x, p_{ij},
$i = 1, n$; $j = 1, \cdots, m$, and with respect to y, q_{ij}, $i = 1, \cdots, n$; $j = 1, m$,
as well as the values, r_{11}, r_{1m}, r_{n1}, and r_{nm} for the mixed partials, must
be supplied by the user.

Description of the parameters:

N,M,NDIM,MDIM,X,Y,U,IR,EPS,IFLAG,
AX,BX,CX,RX,AY,BY,CY,RY as in QUA2D.
P   ARRAY(NDIM,M): Output: Partials with respect to x.
Q   ARRAY(NDIM,M): Output: Partials with respect to y.
R   ARRAY(NDIM,M): Output: Mixed partials.
A   ARRAY(NDIM,MDIM,4,4): Output: Coefficients of the
             bicubic spline interpolant.
DX   ARRAY(N): Work space.
DY   ARRAY(M): Work space.

Required subroutines: TRDISA, CBPERM, TRDISB, CUBMAT.

Figure 4.2. Description of CUB2D.

```
 SUBROUTINE CBPERM(IVJ,M,M1,N,NDIM,N1,N2,P,U,AX,BX,CX,
+                  DX,RX)
 DIMENSION P(NDIM,M),U(NDIM,M),AX(N),BX(N),CX(N),DX(N),
+           RX(N)
 DO 40 J=1,M,M1
     DO 20 I=1,N1
         I1=I-1
         IF(IVJ.NE.0) THEN
             R2=3.*DX(I)*DX(I)*(U(J,I+1)-U(J,I))
             IF(I.EQ.1) GOTO 10
             RX(I1)=R1+R2
             IF(I.EQ.2) RX(I1)=RX(I1)-DX(1)*P(J,1)
             IF(I.EQ.N1) RX(I1)=RX(I1)-DX(N1)*P(J,N)
         ELSE
             R2=3.*DX(I)*DX(I)*(U(I+1,J)-U(I,J))
```

(cont.)

```
                        IF(I.EQ.1) GOTO 10
                        RX(I1)=R1+R2
                        IF(I.EQ.2) RX(I1)=RX(I1)-DX(1)*P(1,J)
                        IF(I.EQ.N1) RX(I1)=RX(I1)-DX(N1)*P(N,J)
                     END IF
10                   R1=R2
20          CONTINUE
            CALL TRDISB(N2,AX,BX,CX,RX)
            DO 30 I=2,N1
                     IF(IVJ.EQ.0) P(I,J)=RX(I-1)
                     IF(IVJ.NE.0) P(J,I)=RX(I-1)
30          CONTINUE
40    CONTINUE
      RETURN
      END
```

Figure 4.3. Program listing of CBPERM.

```
      SUBROUTINE CUBMAT(H,B)
      DIMENSION B(4,4)
      SUM=H*H
      B(3,1)=-3.*SUM
      B(3,2)=-2.*H
      B(3,3)=-B(3,1)
      B(3,4)=-H
      B(4,1)=2.*H*SUM
      B(4,2)=SUM
      B(4,3)=-B(4,1)
      B(4,4)=SUM
      RETURN
      END
```

Figure 4.4. Program listing of CUBMAT.

```
      FUNCTION CBIVAL(U,V,N,NDIM,M,MDIM,X,Y,A,IFLAG)
      DIMENSION X(N),Y(M),A(NDIM,MDIM,4,4)
      DATA I,J/2*1/
      ZERO=0.
      IFLAG=0
      IF(N.LT.2.OR.M.LT.2) THEN
          IFLAG=1
          RETURN
      END IF
      CALL INTTWO(X,N,Y,M,U,V,I,J,IFLAG)
      IF(IFLAG.NE.0) RETURN
      UX=X-X(I)
      VY=V-Y(J)
      CBIVAL=ZERO
      DO 20 K=4,1,-1
          CB=A(I,J,K,4)
          DO 10 L=3,1,-1
              CB=CB*VY+A(I,J,K,L)
10        CONTINUE
          CBIVAL=CBIVAL*UX+CB
20    CONTINUE
      RETURN
      END
```

Figure 4.5. Program listing of CBIVAL.

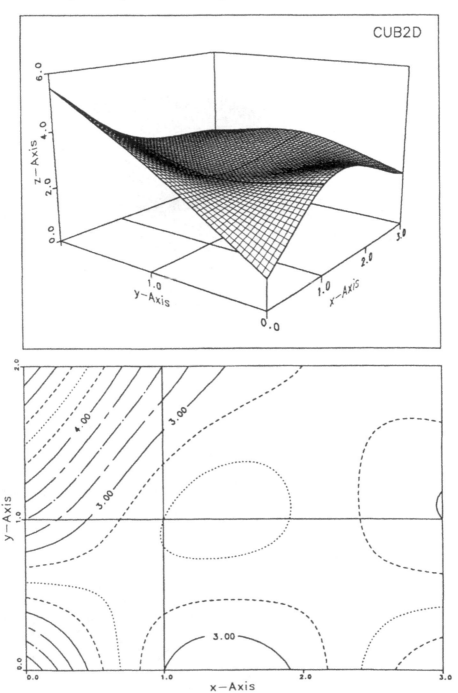

Figure 4.6.

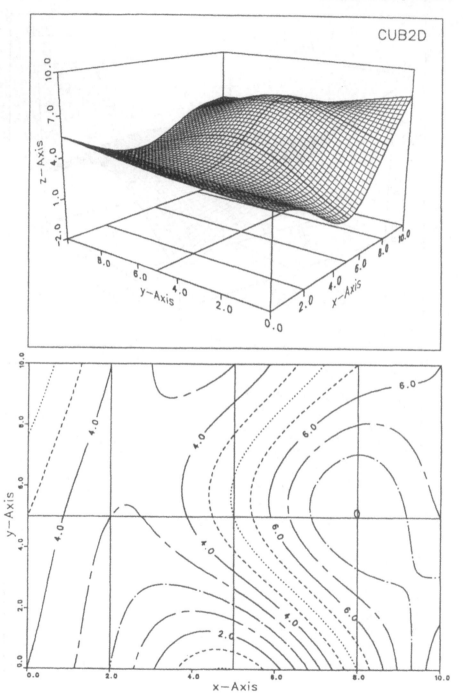

Figure 4.7.

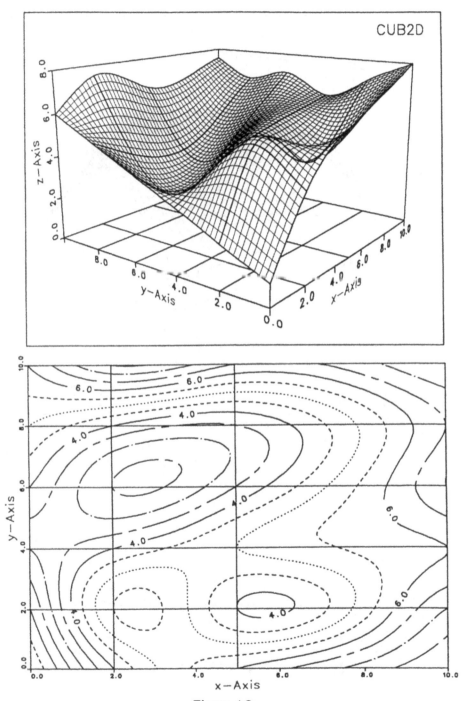

Figure 4.8.

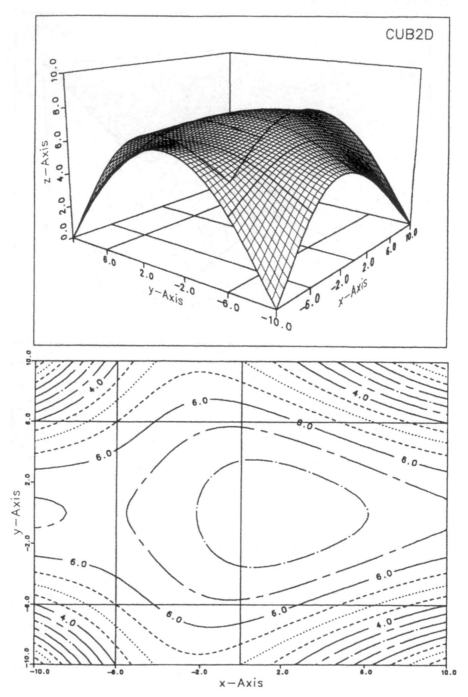

Figure 4.9.

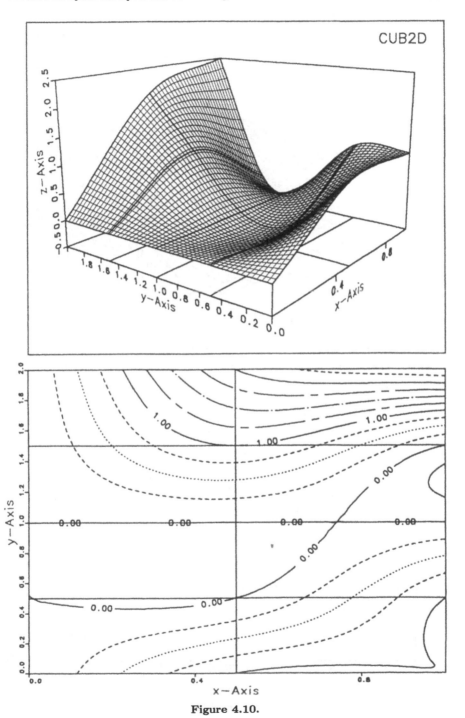

Figure 4.10.

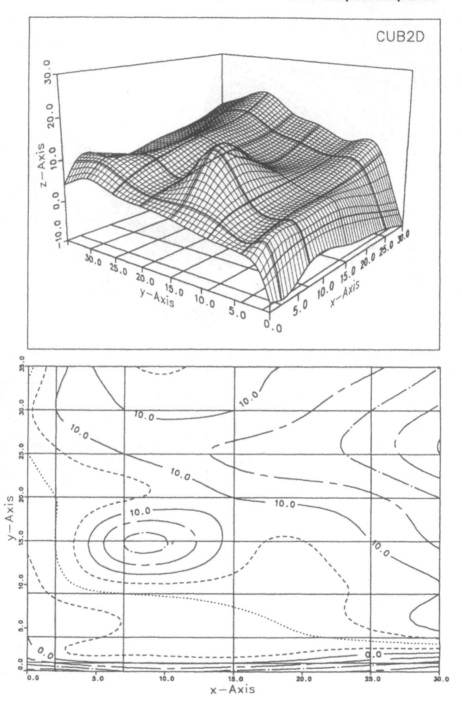

Figure 4.11.

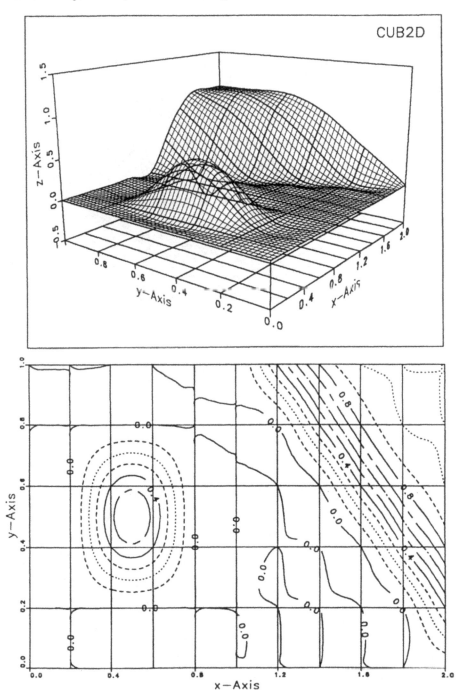

Figure 4.12.

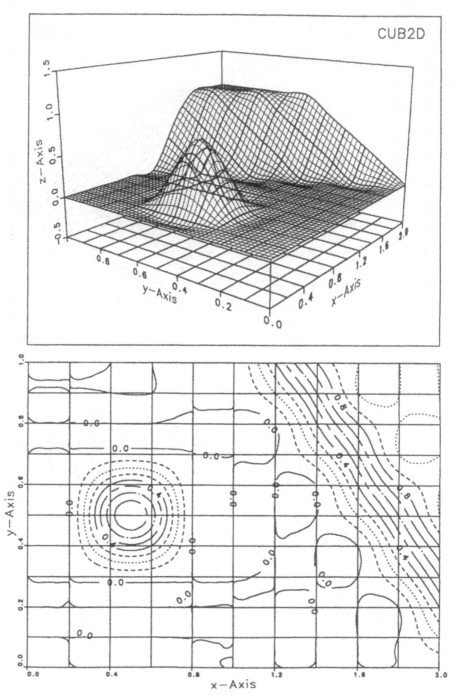

Figure 4.13.

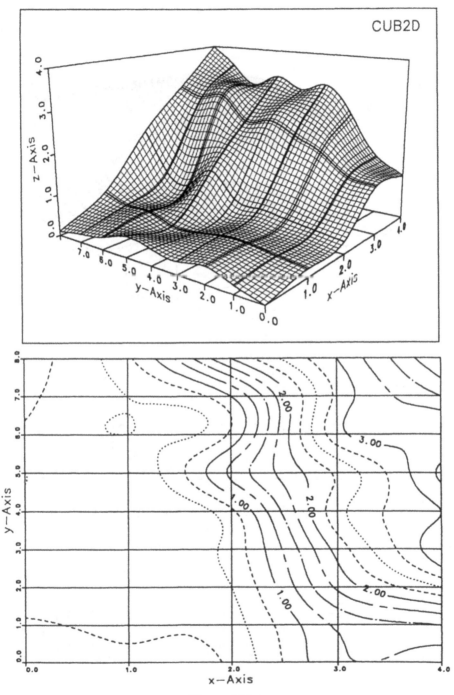

Figure 4.14.

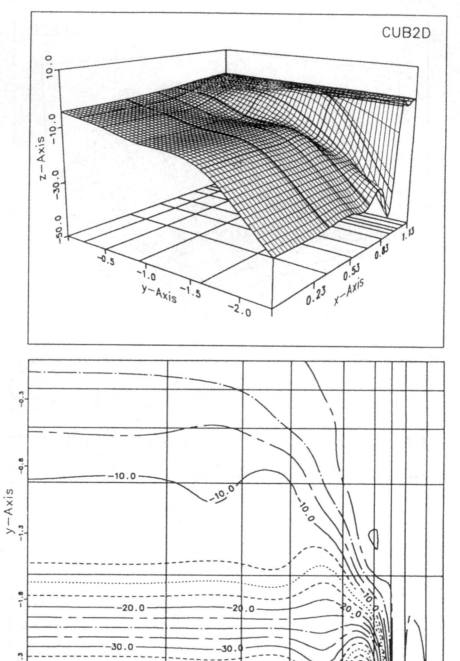

Figure 4.15.

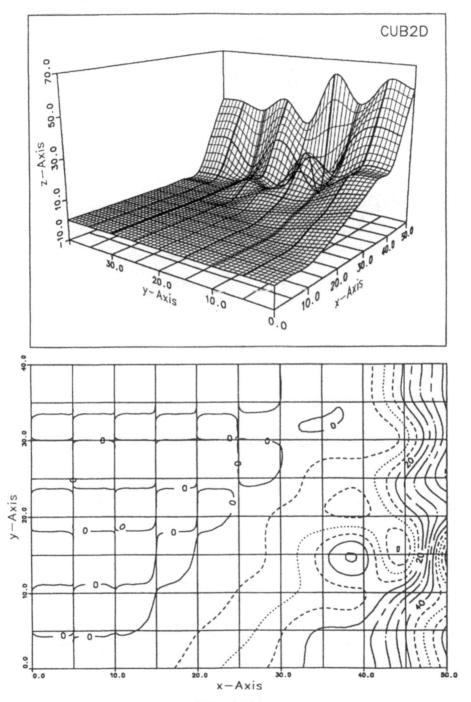

Figure 4.16.

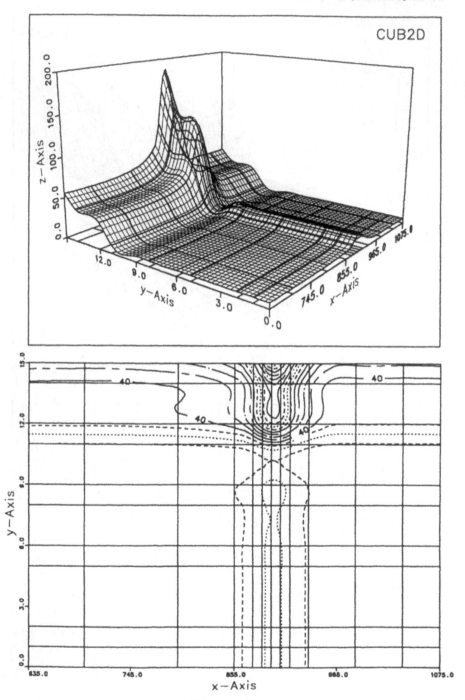

Figure 4.17.

parameters by approximately minimizing, in the least-squares sense, the size of the discontinuity in the partial derivatives $\partial^3 F/\partial x^3$, $\partial^3 F/\partial y^3$, $\partial^4 F/\partial x \partial y^3$, and $\partial^4 F/\partial x^3 \partial y$ at the grid ponts ([107]).

4.2. Parametric Bicubic Spline Interpolation

How to make a good choice of parameters v_k for univariate parametric splines in the plane has recently been a much discussed problem. Since the publication of [100], several important papers have appeared;[1][2] the methods of Foley and Lee have brought about a considerable improvement. In the two-dimensional case, this problem has presumably not yet been investigated to the same extent.

Instead of points in the plane, we start with points in $\mathbb{R}^3$ ([96]),

$$(x_{ij}, y_{ij}, z_{ij}), \quad i = 1, \cdots, n; j = 1, \cdots, m. \tag{4.15}$$

If one imagines that for each $j = 1, \cdots, m$, the points (x_{ij}, y_{ij}, z_{ij}), $i = 1, \cdots, n$, and for each $i = 1, \cdots, n$ the points (x_{ij}, y_{ij}, z_{ij}), $j = 1, \cdots, m$, lie on a curve, then their intersection points would form a non-planar mesh. The $n + m$ curves and the surface on which they lie should, in a certain sense, be as smooth as possible. As one part of a surface may lie above another and it may even intersect itself, a general surface can almost never be represented by a single bivariate function; a parametric representation of the form,

$$\begin{aligned} X &= X(u,v), \\ Y &= Y(u,v), \\ Z &= Z(u,v), \end{aligned} \tag{4.16}$$

becomes necessary. The two parameters, u, and v, will vary in a rectangle in the (u, v)-plane whose corners correspond to the four points (x_{ij}, y_{ij}, z_{ij}), $i = 1, n; j = 1, m$. We then look for grid points (u_i, v_j), $i = 1, \cdots, n; j = 1, \cdots, m$, with

$$0 = u_1 < u_2 < \cdots < u_n = a,$$

$$0 = v_1 < v_2 < \cdots < v_m = b, \tag{4.17}$$

---

[1]Nielson, G.M., and Foley, T.A. "A survey of applicatons of an affine invariant form." In Lyche, T., and Schumaker, L.L., *Mathematical Methods in Computer Aided Geometric Design*, Academic Press, Boston, 1989.

[2]Lee, E.T.Y. "Choosing nodes in parametric curve interpolation," *Computer Aided Design* 21, 363–370 (1989).

so that the surface (4.16) passes through the given points and is as smooth as possible.

"As smooth as possible" we shall define here to mean that we choose a bicubic spline interpolant (IR=1) for each of the three surfaces of (4.16). Hence, we will need to solve three interpolation problems with three different datasets:

$$(u_i, v_j, x_{ij}),$$

$$(u_i, v_j, y_{ij}), \qquad i = 1, \cdots, n; \, j = 1, \cdots, m, \qquad (4.18)$$

$$(u_i, v_j, z_{ij}).$$

Since it is not clear how to choose the u_i and v_j to be optimal in any sense, we will just set (assuming without loss of generality that $a = b = 1$)

$$u_i = \frac{i-1}{n-1}, \qquad i = 1, \cdots, n,$$

$$(4.19)$$

$$v_j = \frac{j-1}{m-1}, \qquad j = 1, \cdots, m.$$

Once this has been done, the coefficient matrices of the three bicubic spline functions, $A = (a_{ijk\ell})$, $B = (b_{ijk\ell})$, and $C = (c_{ijk\ell})$, $i = 1, \cdots, n-1$; $j = 1, \cdots, m-1$; $k, \ell = 1, 2, 3, 4$, may very easily be computed by the following FORTRAN statements:

```
EPS=1E-7
IR=1
CALL CUB2D(N,M,NDIM,MDIM,U,V,X,IR,EPS,P,Q,R,A,IFLAG,
*          DU,AU,BU,CU,RU,DV,AV,BV,CV,RV)
CALL CUB2D(N,M,NDIM,MDIM,U,V,Y,IR,EPS,P,Q,R,B,IFLAG,
*          DU,AU,BU,CU,RU,DV,AV,BV,CV,RV)
CALL CUB2D(N,M,NDIM,MDIM,U,V,Z,IR,EPS,P,Q,R,C,IFLAG,
*          DU,AU,BU,CU,RU,DV,AV,BV,CV,RV)
```

Arbitrary points on the surface so generated can likewise very easily be computed by three consecutive calls to CBIVAL for arbitrary points (v, w) in the rectangle.

Three examples computed in this manner are shown in Figs. 4.18 and 4.19. In Fig. 4.18, for example, the data (4.15) was given by

$$\begin{pmatrix} 0 & 1 \\ 2 & 0 \\ 1 & 1 \end{pmatrix}, \begin{pmatrix} 1 & 0 \\ 0 & 1 \\ 0 & 2 \end{pmatrix}, \begin{pmatrix} 1 & 0 \\ 0 & 0 \\ 1 & 1 \end{pmatrix}.$$

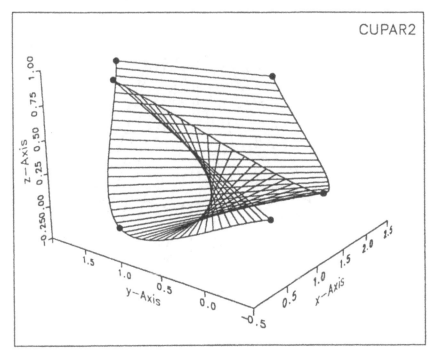

Figure 4.18.

Since we did not implement periodic bicubic spline interpolants, we were not able to close the "lampshade" and the "Möbius band" without them being non-twice continuously differentiable along some line. Clearly, this instrument offers interesting possibilities for surface modeling, at least when C^2 interpolants are desirable. These must still be investigated further. Of course, bilinears, biquadratics, the Hermite cubics of the next section, and birational spline interpolants (which we will cover later) may all be used parametrically in the same simple manner. In these cases also, the problem of how to choose the u_i and v_j is still open. Two local C^1 methods are described in [30, p. 166].

4.3. Bicubic Hermite Spline Interpolation

We now return again to usual bicubic spline interpolation on a rectangular grid by functions of the form (4.1). First, we claim that we can very easily define a local C^0 method on the inner subrectangles R_{ij}, $i = 2, \cdots$, $n - 2$; $j = 2, \cdots, m - 2$. We do this by taking the four corners and their

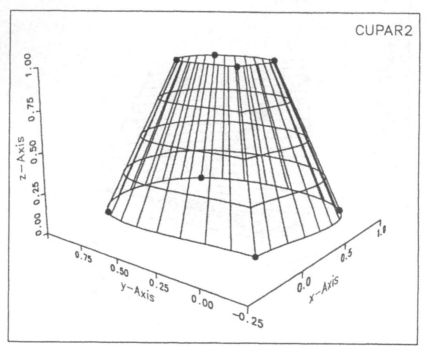

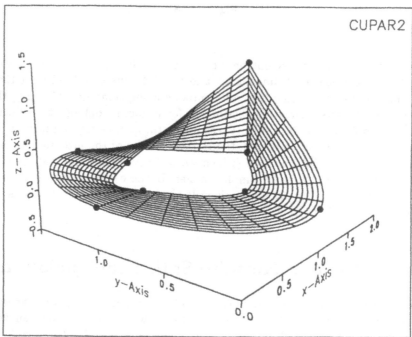

Figure 4.19.

12 neighboring grid points and putting, by means of Lagrange interpolation, a bicubic through these 16 nodes. This interpolant, however, is used for evaluation only in R_{ij}.

A local C^1 method is obtained by simply dropping the C^2 requirements of (4.9), (4.10), (4.13), and (4.14), and either specifying the p, q, and r values arbitrarily or calculating them in some appropriate manner. In this way, the coefficient matrices (4.3) for $i = 1, \cdots, n-1$, and $j = 1, \cdots, m-1$, are still determined. This method gives us a C^1 bicubic Hermite interpolant ([1]). Now F restricts to a cubic polynomial in either x or y on the sides of the subrectangle R_{ij}. Recall that a cubic polynomial is uniquely determined by its function values and first derivatives at two different points. Since the r, p, and q values agree at the corners of the sides, F is continuous across all the sides. Further, F_x is but a cubic in y on $x = x_i$ and F_y is a cubic in x on $y = y_j$. These functions likewise agree on the sides, since they share common p and r or q and r values at the corners. Hence, the interpolant is C^1.

If the required p, q, and r values at the corners of each rectangle are now determined from only nearby data, then we have a *local* bicubic Hermite spline interpolation method. If they were, for example, determined along a grid line only from the given function values on that line, then we would have a *semi-global method*. A large number of such *semi-global* C^1 bicubic Hermite spline interpolation methods are obtained by using the univariate methods and subroutines, GRAD1, GRAD2, GRAD3, GRAD4 ([100]), GRAD2B, GRAD2R, GRAD5, GRAD6, or GRAD7 ([100]). Among these, GRAD5 corresponds to Akima's method, which also has specifically been described ([1,30]) and implemented ([2]) for two dimensions. We have had the best results with GRAD6 and GRAD7. Hence, the subroutine HCUB2D (Figs. 4.20 and 4.21) uses only GRAD7. (In order to use other methods, the user need only replace the call to GRAD7 by another subroutine.)

HCUB2D is not that much different from CUB2D. In fact, the first and last thirds, i.e., the initialization of the parameters and calculation of the coefficients by (4.8), are nearly identical. In the middle part, instead of using (4.9) and (4.10), the p_{ij} are calculated by calling GRAD7 with the x_i and u_{ij}, and similarly the q_{ij} by calling it with the y_j and u_{ij}. Finally, the mixed partials are calculated by *averaging* the results of calls with x_i and q_{ij} and with y_j and p_{ij}. This replaces the use of (4.13) and (4.14). It should be noted that HCUB2D requires $n \geq 3$ and $m \geq 3$.

If you compare the results of HCUB2D (GRAD7) with those of CUB2D, you will see that the difference is hardly noticeable for the examples of Figs. 4.7, 4.8, 4.11, and 4.14. Moreover, in the surfaces corresponding to 4.12, 4.13, and 4.16 there are fewer small waves in the flat part of the interpolating surface. In contrast, Figs. 4.22–4.25 differ considerably, and

```
      SUBROUTINE HCUB2D(N,M,NDIM,MDIM,MAXNM,X,Y,U,EPS,P,Q,R,
     +                  A,IFLAG,DX,DY,AA,BB,CC)
      DIMENSION X(N),Y(M),U(NDIM,M),P(NDIM,M),Q(NDIM,M),
     +          R(NDIM,M),A(NDIM,MDIM,4,4),DX(N),DY(M),
     +          AA(MAXNM),BB(MAXNM),CC(MAXNM),B(4,4),C(4,4),
     +          D(4,4),E(4,4)
      IFLAG=0
      IF(N.LT.3.OR.M.LT.3) THEN
          IFLAG=1
          RETURN
      END IF
      N1=N-1
      N2=N-2
      M1=M-1
      M2=M-2
      ZERO=0.
      DO 10 I=1,N1
         DX(I)=1./(X(I+1)-X(I))
10    CONTINUE
      DO 20 J=1,M1
         DY(J)=1./(Y(J+1)-Y(J))
20    CONTINUE
      DO 40 I=1,4
         DO 30 J=1,4
            B(I,J)=ZERO
            E(I,J)=ZERO
30       CONTINUE
40    CONTINUE
      B(1,1)=1.
      B(2,2)=1.
      E(1,1)=1.
      E(2,2)=1.
      DO 70 J=1,M
         DO 50 I=1,N
            AA(I)=U(I,J)
50       CONTINUE
         CALL GRAD7(N,X,AA,EPS,BB,IFLAG,CC)
         DO 60 I=1,N
            P(I,J)=BB(I)
60       CONTINUE
70    CONTINUE
      DO 100 I=1,N
         DO 80 J=1,M
            AA(J)=U(I,J)
80       CONTINUE
         CALL GRAD7(M,Y,AA,EPS,BB,IFLAG,CC)
         DO 90 J=1,M
            Q(I,J)=BB(J)
90       CONTINUE
100   CONTINUE
      DO 130 J=1,M
         DO 110 I=1,N
            AA(I)=Q(I,J)
110      CONTINUE
         CALL GRAD7(N,X,AA,EPS,BB,IFLAG,CC)
         DO 120 I=1,N
            R(I,J)=BB(I)
```

(cont.)

```
120     CONTINUE
130 CONTINUE
    DO 160 I=1,N
        DO 140 J=1,M
            AA(J)=P(I,J)
140     CONTINUE
        CALL GRAD7(M,Y,AA,EPS,BB,IFLAG,CC)
        DO 150 J=1,M
            R(I,J)=(R(I,J)+BB(J))/2.
150     CONTINUE
160 CONTINUE
    DO 240 I=1,N1
        I1=I+1
        CALL CUBMAT(DX(I),B)
        DO 230 J=1,M1
            J1=J+1
            C(1,1)=U(I,J)
            C(1,2)=Q(I,J)
            C(2,1)=P(I,J)
            C(2,2)=R(I,J)
            C(1,3)=U(I,J1)
            C(1,4)=Q(I,J1)
            C(2,3)=P(I,J1)
            C(2,4)=R(I,J1)
            C(3,1)=U(I1,J)
            C(3,2)=Q(I1,J)
            C(4,1)=P(I1,J)
            C(4,2)=R(I1,J)
            C(3,3)=U(I1,J1)
            C(3,4)=Q(I1,J1)
            C(4,3)=P(I1,J1)
            C(4,4)=R(I1,J1)
            DO 190 K1=1,4
                DO 180 K2=1,4
                    SUM=ZERO
                    DO 170 K=1,4
                        SUM=SUM+B(K1,K)*C(K,K2)
170                 CONTINUE
                    D(K1,K2)=SUM
180             CONTINUE
190         CONTINUE
            CALL CUBMAT(DY(J),E)
            DO 220 K1=1,4
                DO 210 K2=1,4
                    SUM=ZERO
                    DO 200 K=1,4
                        SUM=SUM+D(K1,K)*E(K2,K)
200                 CONTINUE
                    A(I,J,K1,K2)=SUM
210             CONTINUE
220         CONTINUE
230     CONTINUE
240 CONTINUE
    RETURN
    END
```

Figure 4.20. Program listing of HCUB2D.

Calling sequence:

CALL HCUB2D(N,M,NDIM,MDIM,MAXNM,X,Y,U,EPS,P,Q,R,
           A,IFLAG,DX,DY,AA,BB,CC)

Purpose:

For given (x_i, y_j) $i = 1, \cdots, n \geq 2$; $j = 1, \cdots, m \geq 2$, in the plane with $x_1 < \cdots < x_n$ and $y_1 < \cdots < y_m$ and associated heights u_{ij}, this routine determines the coefficients $a_{ijk\ell}$ of a bicubic Hermite spline interpolant, F, i.e., F is of the form (1.12) with $N = 4$ on each of the subrectangles R_{ij}. Values for the partial and mixed partial derivatives are computed using the subroutine GRAD7 ([100]).

Description of the parameters:

N,M,NDIM,MDIM,X,Y,U,P,Q,R,A,DX,DY as in CUB2D.
MAXNM   Maximum of N and M.
EPS   Parameter used in calling GRAD7. Recommendation: EPS= 10^{-t}, where t denotes the number of available decimal digits. If the absolute value of a number is smaller than EPS, it is interpreted as zero.
IFLAG   =0:   Normal execution.
        =1:   N<3 or M<3.
AA,BB,CC ARRAY(MAXNM): Work space.

Required subroutines: GRAD7, CUBMAT.

Figure 4.21. Description of HCUB2D.

indeed in a positive sense, from 4.9, 4.10, 4.15, and 4.17. Figures 4.24 and 4.25 have a very nice appearance, but a close inspection reveals that there is still room for improvement (by later methods).

Replacing GRAD7 by GRAD5(ICASE=2) has very little effect on the figures (this is also true for further examples). Even if the so-called *twist* values r_{ij} are all simply set to zero, this has only a minor effect on these two variants of HCUB2D.

The subroutine ITPLBV([2]), in comparison to Figs. 4.22 and 4.25, gives much worse results. For Fig. 4.24 the results were comparable, and for 4.23 even somewhat better. A slightly modified version is available on the diskette of subroutines, but we do not list it here. In order to eliminate unwanted oscillations in C^2 bicubic spline interpolants, it is suggested in [10] to locally adjust the resulting *twist* values (and only these) in some appropriate fashion.

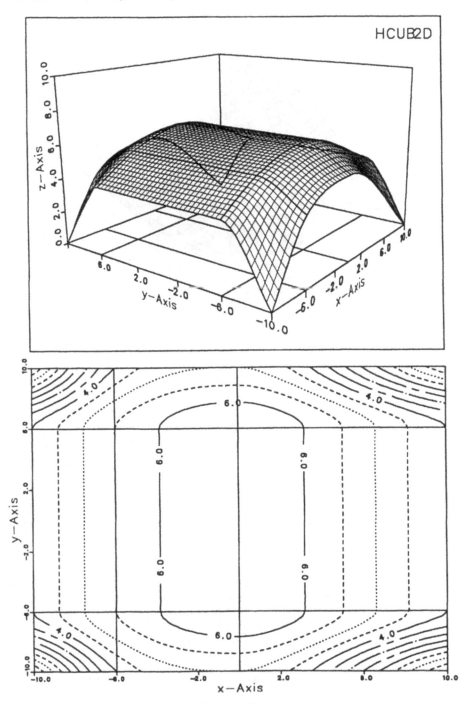

Figure 4.22.

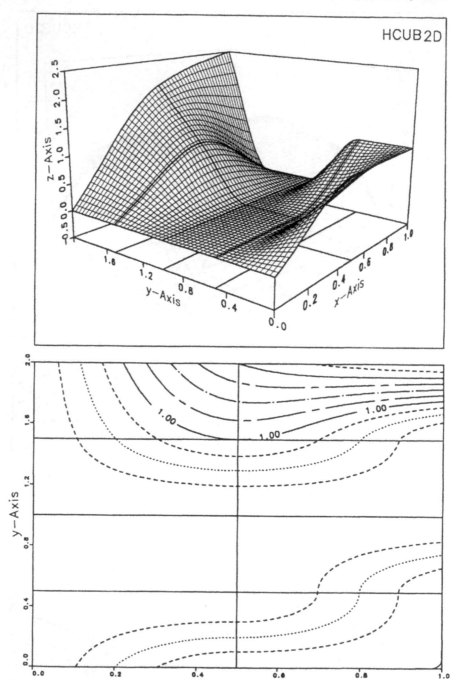

Figure 4.23.

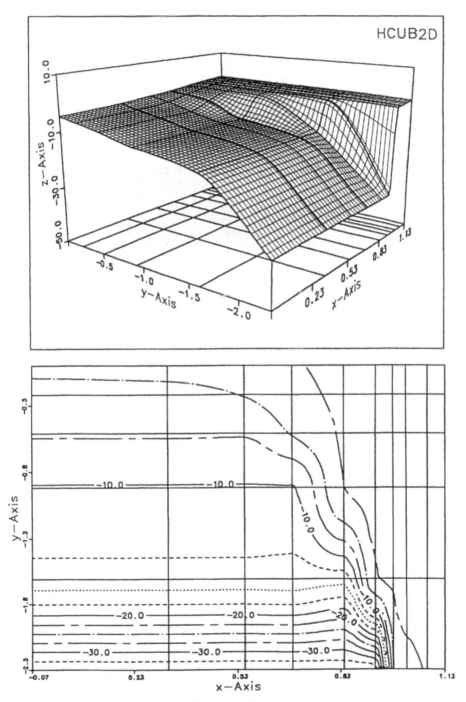

Figure 4.24.

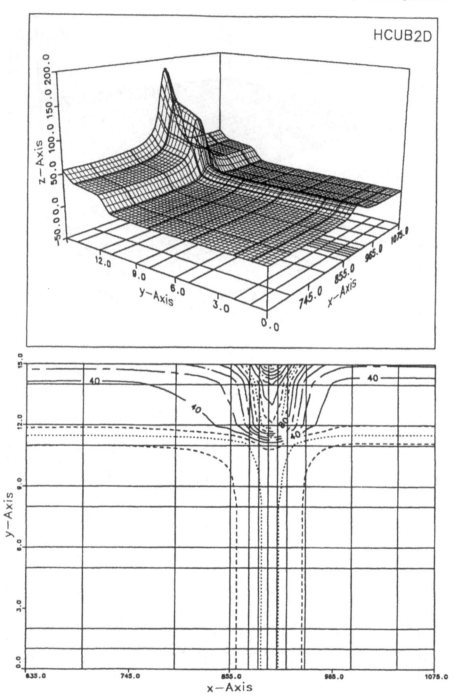

Figure 4.25.

4.4. Semi-Bicubic Hermite Spline Interpolation

In what follows, we describe a method of Hermite interpolation that does not require any estimates of the twists r_{ij}. This is accomplished by eliminating the monomials, s^2t^2, s^3t^2, s^2t^3, and s^3t^3 (semi-bicubics), from the form (4.1) and, with

$$s = x - x_i, \quad t = y - y_j, \tag{4.20}$$

setting

$$
\begin{aligned}
G(x,y) \;=\; & a_1 + a_2 s + s_3 t + a_4 st \\
& + s^2(b_1 + b_2 s + b_3 t + b_4 st) \\
& + t^2(c_1 + c_2 s + c_3 t + c_4 st)
\end{aligned}
\tag{4.21}
$$

on the rectangle R_{ij}. The 12, parameters a_1, a_2, a_3, a_4, b_1, b_2, b_3, b_4, c_1, c_2, c_3, and c_4 that need to be determined then match in number the total of 12 interpolation conditions for the u, p, and q values at the corners of R_{ij}. We will show that this system is uniquely solvable. The global function will only be *semi-C^1*, meaning that only $\partial G/\partial x$ or $\partial G/\partial y$ is continuous across a given grid line.

Since

$$
\begin{aligned}
\frac{\partial G}{\partial x} \;=\; & a_2 + a_4 t + 2s(b_1 + b_2 s + b_3 t + b_4 st) \\
& + s^2(b_2 + b_4 t) + t^2(c_2 + c_4 t),
\end{aligned}
$$

$$
\begin{aligned}
\frac{\partial G}{\partial y} \;=\; & a_3 + a_4 s + 2t(c_1 + c_2 s + c_3 t + c_4 st) \\
& + s^2(b_3 + b_4 s) + t^2(c_3 + c_4 s),
\end{aligned}
$$

it follows that

$$G(x_i, y_j) = u_{ij} = a_1, \tag{4.22}$$

$$G(x_{i+1}, y_j) = u_{i+1,j} = a_1 + \Delta x_i a_2 + \Delta x_i^2(b_1 + \Delta x_i b_2), \tag{4.23}$$

$$
\begin{aligned}
G(x_{i+1}, y_{j+1}) = u_{i+1,j+1} = \; & a_1 + \Delta x_i a_2 + \Delta y_j a_3 + \Delta x_i \Delta y_j a_4 \\
& + \Delta x_i^2(b_1 + \Delta x_i b_2 + \Delta y_j b_3 + \Delta x_i \Delta y_j b_4) \\
& + \Delta y_j^2(c_1 + \Delta x_i c_2 + \Delta y_j c_3 + \Delta x_i \Delta y_j c_4),
\end{aligned}
\tag{4.24}
$$

$$G(x_i, y_{j+1}) = u_{i,j+1} = a_1 + \Delta y_j a_3 + \Delta y_j^2(c_1 + \Delta y_j c_3), \tag{4.25}$$

$$\frac{\partial G}{\partial x}(x_i, y_j) = p_{ij} = a_2, \tag{4.26}$$

$$\frac{\partial G}{\partial x}(x_{i+1}, y_j) = p_{i+1,j} = a_2 + 2\Delta x_i(b_1 + \Delta x_i b_2) + \Delta x_i^2 b_2, \quad (4.27)$$

$$\frac{\partial G}{\partial x}(x_{i+1}, y_{j+1}) = p_{i+1,j+1} = a_2 + \Delta y_j a_4$$
$$+2\Delta x_i(b_1 + \Delta x_i b_2 + \Delta y_j b_3 + \Delta x_i \Delta y_j b_4)$$
$$+\Delta x_i^2(b_2 + \Delta y_j b_4) \quad (4.28)$$
$$+\Delta y_j^2(c_2 + \Delta y_j c_4),$$

$$\frac{\partial G}{\partial x}(x_i, y_{j+1}) = p_{i,j+1} = a_2 + \Delta y_j a_4 + \Delta y_j^2(c_2 + \Delta y_j c_4), \quad (4.29)$$

$$\frac{\partial G}{\partial y}(x_i, y_j) = q_{ij} = a_3, \quad (4.30)$$

$$\frac{\partial G}{\partial y}(x_{i+1}, y_j) = q_{i+1,j} = a_3 + \Delta x_i a_4 + \Delta x_i^2(b_3 + \Delta x_i b_4), \quad (4.31)$$

$$\frac{\partial G}{\partial y}(x_{i+1}, y_{j+1}) = q_{i+1,j+1} = a_3 + \Delta x_i a_4,$$
$$+2\Delta y_j(c_1 + \Delta x_i c_2 + \Delta y_j c_3 + \Delta x_i \Delta y_j c_4)$$
$$+\Delta x_i^2(b_3 + \Delta x_i b_4) \quad (4.32)$$
$$+\Delta y_j^2(c_3 + \Delta x_i c_4),$$

$$\frac{\partial G}{\partial y}(x_i, y_{j+1}) = q_{i,j+1} = a_3 + 2\Delta y_j(c_1 + \Delta y_j c_3) + \Delta y_j^2 c_3. \quad (4.33)$$

From (4.22), (4.26), and (4.30), we immediately have

$$a_1 = u_{ij},\ a_2 = p_{ij},\ a_3 = q_{ij}. \quad (4.34)$$

Equations (4.23) and (4.27) can then be written in the form,

$$\Delta x_i^2 b_1 + \Delta x_i^3 b_2 = r_1 := u_{i+1,j} - a_1 - \Delta x_i a_2,$$
$$2\Delta x_i b_1 + 3\Delta x_i^3 b_2 = r_2 := p_{i+1,j} - a_2,$$

from which it follows that

$$b_1 = \frac{1}{\Delta x_i^2}(3r_1 - \Delta x_i r_2),\ b_2 = \frac{1}{\Delta x_i^2}(-\frac{2}{\Delta x_i}r_1 + r_2). \quad (4.35)$$

Similarly, from (4.25) and (4.33), it follows that

$$c_1 = \frac{1}{\Delta y_j^2}(3r_3 - \Delta y_j r_4),\ c_3 = \frac{1}{\Delta y_j^2}(-\frac{2}{\Delta y_j}r_3 + r_4), \quad (4.36)$$

where

$$r_3 = u_{i,j+1} - a_1 - \Delta y_j a_3$$

and
$$r_4 = q_{i,j+1} - a_3.$$

Equations (4.24), (4.28), (4.29), (4.31), and (4.32), which have not yet been used, may be reexpressed by moving the already calculated coefficients to the right side. This results in

$$\Delta y_j a_4 + \Delta x_i \Delta y_j b_3 + \Delta x_i^2 \Delta y_j b_4 + \Delta y_j^2 c_2 + \Delta y_j^3 c_4 = r_5, \quad (4.37)$$

$$\Delta y_j a_4 + 2\Delta x_i \Delta y_j b_3 + 3\Delta x_i^2 \Delta y_j b_4 + \Delta y_j^2 c_2 + \Delta y_j^3 c_4 = r_6, \quad (4.38)$$

$$\Delta y_j a_4 + \Delta y_j^2 c_2 + \Delta y_j^3 c_4 = r_7, \quad (4.39)$$

$$\Delta x_i a_4 + \Delta x_i^2 b_3 + \Delta x_i^3 b_4 = r_8, \quad (4.40)$$

$$\Delta x_i a_4 + \Delta x_i^2 b_3 + \Delta x_i^3 b_4 + 2\Delta x_i \Delta y_j c_2 + 3\Delta x_i \Delta y_j^2 c_4 = r_9, \quad (4.41)$$

where

$$r_5 = (u_{i+1,j+1} - a_1 - \Delta x_i a_2 - \Delta y_j a_3 - \Delta x_i^2 b_1 - \Delta x_i^3 b_2$$
$$\Delta y_j^2 c_1 - \Delta y_j^3 c_0)/\Delta r_i,$$

$$r_6 = p_{i+1,j+1} - a_2 - 2\Delta x_i b_1 - 3\Delta x_i^2 b_2,$$

$$r_7 = p_{i,j+1} - a_2,$$

$$r_8 = q_{i+1,j} - a_3,$$

$$r_9 = q_{i+1,j+1} - a_3 - 2\Delta y_j c_1 - 3\Delta y_j^2 c_3.$$

From the differences of (4.37) and (4.39) and of (4.38) and (4.39), we obtain

$$b_3 = \frac{1}{\Delta x_i}(3r_{10} - r_{11}), \quad (4.42)$$

$$b_4 = \frac{1}{\Delta x_i^2}(-2r_{10} + r_{11}), \quad (4.43)$$

where

$$r_{10} = \frac{r_5 - r_7}{\Delta y_j},$$

$$r_{11} = \frac{r_6 - r_7}{\Delta y_j}.$$

Then from (4.40),

$$a_4 = \frac{1}{\Delta x_i}(r_8 - \Delta x_i^2 b_3 - \Delta x_i^3 b_4). \quad (4.44)$$

Finally, from equations (4.39) and (4.41), we may compute

$$c_2 = 3r_{12} - r_{13}, \quad (4.45)$$

$$c_4 = \frac{1}{\Delta y_j}(-2r_{12} + r_{13}), \quad (4.46)$$

where we have set

$$r_{12} = \frac{1}{\Delta y_j^2}(r_7 - \Delta y_j a_4),$$

$$r_{13} = \frac{1}{\Delta x_i \Delta y_j}(r_9 - \Delta x_i a_4 - \Delta x_i^2 b_3 - \Delta x_i^3 b_4).$$

By clever eliminations, we reduced our problem to solving a number of 2×2 systems. All the desired coefficients are now at hand. It remains to be seen by example how semi-bicubic Hermite compares to bicubic Hermite and biquadratic Hermite C^1 spline interpolation. For the latter, each rectangle is divided into four subrectangles, each with its own biquadratic function. For reasons of space, we have omitted this comparison.

4.5. Shape Preservation

In early survey articles ([30,93]), the problem of shape preservation is not yet mentioned at all. In practical problems, where the given data, and in particular their number, are generally fixed, we always look for shape preserving (piecewise positive, monotone, convex etc.) and visually pleasing curves and surfaces. Analytic results on the rates of convergence of various interpolation methods as the number of grid points increases are almost never very interesting in this context. In the univariate case, shape preservation has already been comprehensively studied (see the references of [100]). For two dimensions, there are a small number of publications on this topic ([11,12,13,18,19,20,23,44,90,105]). We will make some brief references to these in what follows. Programs, other than BIMOND1([12]), the newer version BIMOND3([45]), and SPBSI1([20]), are almost nonexistent.

In order to reduce the number of undesirable local extrema and saddle points (similar to the method of [1]), it is suggested in [23] that one use the analysis behind GRAD6([100]) together with C^1 quadratic splines to determine function values and first partial derivative values on the sides of R, to set $r_{ij} = 0$, and then finally to make use of Hermite interpolation. This, however, only guarantees shape preservation on the boundary.

Sufficient conditions on the first and mixed partial derivatives at the grid points for the bicubic Hermite interpolant of *bimonotone* data to be a *bimonotone* function are discussed in [12] and improved upon in [11,13]. Here, bimonotone data is data that is either always increasing or always decreasing along all grid lines. A bimonotone function, F, is one for which for all x, $F(x, \cdot)$, and for all y, $F(\cdot, y)$, are always decreasing or always increasing. The data of Figs. 2.11, 4.15, and 4.24 are taken from [12]. Corresponding programs, BIMOND1([12]) and BIMOND3([45]), are available.

There are a total of $22mn - 18m - 18n + 12$ linear inequalities that the parameters p_{ij}, q_{ij}, and r_{ij} must satisfy. Since there is no suitable objective function known, the algorithm only finds some solution satisfying these inequalities, and this at great expense. To be sure, BIMOND1 does find a good solution for the data of Fig. 4.24. It is given in [11,12].

Instead of proceeding from bicubics and determining the p_{ij}, q_{ij}, and r_{ij} so as to have shape-preserving properties (here monotonicity), [19] takes a completely different approach. It is assumed that these values are either known or have been calculated in some appropriate manner. Then, on each rectangle, exponents $0 \leq n_k, m_\ell < \infty$ in

$$G(x,y) = \sum_{k=1}^{4} \sum_{\ell=1}^{4} b_{ijk\ell}(x - x_i)^{n_k}(y - y_j)^{m_\ell} \qquad (4.47)$$

are determined (and subsequently also the coefficients $b_{ijk\ell}$ from the u_{ij}, p_{ij}, q_{ij}, and r_{ij}) so that the global function, F, defined on all of R, is s ($s \geq 1$) times continuously differentiable and so that the $F_{ij} = G$ are either monotone or convex (concave) or even that they have no particular shape preserving properties. The method is *semi-local*, as the data from rectangles R_{ij}, $j = 1, \cdots, m-1$, and R_{ij}, $i = 1, \cdots, n-1$, are involved in the determination of the n_k and m_ℓ. To be more precise, Bernstein polynomials are used instead of the above form. A disadvantage is that very large exponents n_k and m_ℓ often are required. The solutions for the examples of Figs. 4.15, 4.16, and 4.17 ([18,19]) are convincing. A FORTRAN-77 program SPBSI1 has been announced in [20].

Reference [90] considers *positive interpolation* of positive data u_{ij} and *convex interpolation* of convex data on rectangular grids using, especially, cubic Hermite splines. If (4.1) is written in a more suitable form, namely, in terms of nonnegative cubics g_{ik}, $k = 1, 2, 3, 4$, then it can be immediately seen for $u_{ij} \geq 0$ that, for example, zero values for the p, q, and r values at all the corners of R_{ij} forces the nonnegativity of G on R_{ij}. However, this method forces the tangent planes at the four grid points to be parallel to the xy-plane, and so it rarely produces visually appealing interpolation surfaces. Certain nonlinear inequalities on the p, q, and r values at the four grid points result when G is written in the form for applying bivariate Horner's rule and then asking that the various univariate cubics in x multiplying $y^{\ell-1}$, $\ell = 1, \cdots, 4$, and those in y multiplying x^{k-1}, $k = 1, \cdots, 4$, be nonnegative. Although it is not possible to solve this system of nonlinear inequalities, one can always check to see if they are satisfied for any particular given values.

In exactly the same way, there is a condition on the p, q, and r values at the corners of a subrectangle R_{ij} that checks whether the convexity conditions,

$$\frac{\partial^2 G}{\partial x^2} \geq 0, \ \frac{\partial^2 G}{\partial y^2} \geq 0, \ \text{and} \ \frac{\partial^2 G}{\partial x^2}\frac{\partial^2 G}{\partial y^2} - \left(\frac{\partial^2 G}{\partial x \partial y}\right)^2 \geq 0, \qquad (4.48)$$

hold in the interior. This condition is also computationally intractable. Somewhat simpler conditions result when it is only asked that G be s-convex, i.e., for each fixed x the functions $G(x, \cdot)$ and for each fixed y the functions $G(\cdot, y)$ are convex. Altogether, the situation is rather hopeless, since it is possible to have convex data that can be interpolated neither by convex nor by s-convex bicubic Hermite splines ([90]).

Analogous questions, applying univariate results, for biquadratic splines with knots the same as nodes and for certain (quadratic) birational splines are treated in [90]. For these, the corresponding systems of inequalities are somewhat simpler, although they are still nonlinear.

4.6. Biquadratic Histosplines

Univariate histosplines ([100]) can be generalized to two variables ([66,104]). We will still refer to this generalization as *histosplines*.

We are now given rectangular boxes of dimensions $\Delta x_i \times \Delta y_j \times v_{ij}$, above each subrectangle R_{ij}, $i = 1, \cdots, n-1; j = 1, \cdots, m-1$. v_{ij} is then the height of the box above R_{ij}, and so we also refer to it as the *volume height*. We look for biquadratic functions $F_{ij} = G$ of the form (3.1) on each subrectangle so that the resulting global function, F, defined on all of R, has (to start with) continuous first partials across the grid lines and also satisfies the volume conditions,

$$\frac{1}{\Delta x_i \Delta y_j} \int_{x_i}^{x_{i+1}} \int_{y_j}^{y_{j+1}} F(x,y)dydx = v_{ij}, \quad i = 1, \cdots, n-1; j = 1, \cdots, m-1.$$
$$(4.49)$$

It turns out that F is uniquely determined given certain boundary values.

First, we show how in analogy to the univariate case ([100]) we can use C^2 bicubic splines to easily find the required C^1 histosplines ([102]). Assuming that we are given the boundary values (4.11) and (4.12), we set

$$
\begin{aligned}
u_{1j} &= 0, \quad j = 1, \cdots, m, \\
u_{i1} &= 0, \quad i = 1, \cdots, n, \\
u_{ij} &= \sum_{k=1}^{i-1}\sum_{\ell=1}^{j-1} \Delta x_k \Delta y_\ell v_{k\ell}, \quad i = 2, \cdots, n; j = 2, \cdots, m,
\end{aligned}
$$
$$(4.50)$$

and calculate the C^2 bicubic spline interpolant, $\overline{F}$, of this data. Then

$$F = \overline{F}_{xy} \tag{4.51}$$

has the desired properties.

In fact, by (4.51), F is a C^1 biquadratic spline. Moreover, $F = \overline{F}_{xy}$ satisfies the volume conditions (4.49), since by (4.50),

$$
\begin{aligned}
&\int_{x_i}^{x_{i+1}} \int_{y_j}^{y_{j+1}} \overline{F}_{xy}(x,y)\,dy\,dx \\
&= \int_{x_i}^{x_{i+1}} \left[\overline{F}_x(x,y)\Big|_{y_j}^{y_{j+1}}\right] dx \\
&= \int_{x_i}^{x_{i+1}} \overline{F}_x(x,y_{j+1})\,dx - \int_{x_i}^{x_{i+1}} \overline{F}_x(x,y_j)\,dx \\
&= \overline{F}(x,y_{j+1})\Big|_{x_i}^{x_{i+1}} - \overline{F}(x,y_j)\Big|_{x_i}^{x_{i+1}} \\
&= (u_{i+1,j+1} - u_{i,j+1}) - (u_{i+1,j} - u_{ij}) \\
&= \left(\sum_{k=1}^{i}\sum_{\ell=1}^{j} \Delta x_k \Delta y_\ell v_{k\ell} - \sum_{k=1}^{i-1}\sum_{\ell=1}^{j} \Delta x_k \Delta y_\ell v_{k\ell}\right) \\
&\quad - \left(\sum_{k=1}^{i}\sum_{\ell=1}^{j-1} \Delta x_k \Delta y_\ell v_{k\ell} - \sum_{k=1}^{i-1}\sum_{\ell=1}^{j-1} \Delta x_k \Delta y_\ell v_{k\ell}\right) \\
&= \sum_{\ell=1}^{j} \Delta x_i \Delta y_\ell v_{i\ell} - \sum_{\ell=1}^{j-1} \Delta x_i \Delta y_\ell v_{i\ell} \\
&= \Delta x_i \Delta y_j v_{ij}.
\end{aligned}
$$

(In the preceding manipulations, we followed the usual summation convention of, for example, interpreting $\sum_{k=1}^{0}$ as zero.)

By (4.51), the coefficients $a_{ijk\ell}$ of F, written in the form (3.1), are very easily obtained from those of $\overline{F}$. Denoting these latter by $\overline{a}_{ijk\ell}$, we have

$$a_{ijk\ell} = k \cdot \ell \cdot \overline{a}_{i,j,k+1,\ell+1}, \quad k,\ell = 1,2,3. \tag{4.52}$$

Now we must still clarify what are reasonable values for the boundary values (4.11) and (4.12) that are to be passed to CUB2D when calculating histosplines. As these amount in general to specifying values for univariate integrals along the boundary ([102]), it is not clear how to do this. Hence, we wish to work out explicitly the simple difference quotients used in CUB2D with IR=1. These are

$$p_{1j} = \frac{u_{2j} - u_{1j}}{\Delta x_1}, \quad j = 1,\cdots,m,$$

$$p_{nj} = \frac{u_{nj} - u_{n-1,j}}{\Delta x_{n-1}}, \quad j = 1, \cdots, m, \tag{4.53}$$

$$q_{i1} = \frac{u_{i2} - u_{i1}}{\Delta y_1}, \quad i = 1, \cdots, n,$$

$$q_{im} = \frac{u_{im} - u_{i,m-1}}{\Delta y_{m-1}}, \quad i = 1, \cdots, n.$$

For the specific values of (4.50), these become

$$p_{11} = 0, \quad p_{1j} = \sum_{\ell=1}^{j-1} \Delta y_\ell v_{1\ell}, \quad j = 2, \cdots, m,$$

$$p_{n1} = 0, \quad p_{nj} = \sum_{\ell=1}^{j-1} \Delta y_\ell v_{n-1,\ell}, \quad j = 2, \cdots, m,$$

$$q_{11} = 0, \quad q_{i1} = \sum_{k=1}^{i-1} \Delta x_k v_{k1}, \quad i = 2, \cdots, n, \tag{4.54}$$

$$q_{1m} = 0, \quad q_{im} = \sum_{k=1}^{i-1} \Delta x_k v_{k,m-1}, \quad i = 2, \cdots, n.$$

Furthermore, the mixed partials are set by CUB2D to

$$r_{11} = \frac{1}{2}\left(\frac{p_{12} - p_{11}}{\Delta y_1} + \frac{q_{21} - q_{11}}{\Delta x_1}\right),$$

$$r_{n1} = \frac{1}{2}\left(\frac{p_{n2} - p_{n1}}{\Delta y_1} + \frac{q_{n1} - q_{n-1,1}}{\Delta x_{n-1}}\right), \tag{4.55}$$

$$r_{nm} = \frac{1}{2}\left(\frac{p_{nm} - p_{n,m-1}}{\Delta y_{m-1}} + \frac{q_{nm} - q_{n-1,m}}{\Delta x_{n-1}}\right),$$

$$r_{1m} = \frac{1}{2}\left(\frac{p_{1m} - p_{1,m-1}}{\Delta y_{m-1}} + \frac{q_{2m} - q_{1m}}{\Delta x_1}\right),$$

which for the specific values of (4.50) become

$$r_{11} = v_{11},$$
$$r_{n1} = v_{n-1,1}, \tag{4.56}$$
$$r_{nm} = v_{n-1,m-1},$$
$$r_{1m} = v_{1,m-1}.$$

Formula (4.56) implies that at the corners of R the corresponding volume heights are specified as function values for F. Since this need not always be the best choice, we have arranged that in the implementation of this method, the user has the option of specifying other values, even if he would still like to use (4.54).

```
      SUBROUTINE QVASPL(N,M,NDIM,MDIM,X,Y,V,U,IR,EPS,P,Q,R,A,
     +           IFLAG,DX,AX,BX,CX,RX,DY,AY,BY,CY,RY)
      DIMENSION X(N),Y(M),V(NDIM,M),U(NDIM,M),P(NDIM,M),
     +          Q(NDIM,M),R(NDIM,M),A(NDIM,MDIM,4,4),
     +          DX(N),AX(N),BX(N),CX(N),RX(N),
     +          DY(M),AY(M),BY(M),CY(M),RY(M)
      IFLAG=0
      IF(N.LT.2.OR.M.LT.2) THEN
          IFLAG=1
          RETURN
      END IF
      N1=N-1
      M1=M-1
      ZERO=0.
      DO 10 I=1,N1
          DX(I)=X(I+1)-X(I)
          U(I,1)=ZERO
10    CONTINUE
      DO 20 J=1,M1
          DY(J)=Y(J+1)-Y(J)
          U(1,J)=ZERO
20    CONTINUE
      U(N,1)=ZERO
      U(1,M)=ZERO
      DO 50 I=2,N
          DO 40 J=2,M
              H=ZERO
              J1=J-1
              DO 30 K=1,I-1
                  H=H+DX(K)*DY(J1)*V(K,J1)
30            CONTINUE
              U(I,J)=U(I,J1)+H
40        CONTINUE
50    CONTINUE
      IF(IR.EQ.1) GOTO 80
      Q(1,1)=ZERO
      Q(1,M)=ZERO
      DO 60 I=2,N
          I1=I-1
          Q(I,1)=Q(I1,1)+DX(I1)*V(I1,1)
          Q(I,M)=Q(I1,M)+DX(I1)*V(I1,M1)
60    CONTINUE
      P(1,1)=ZERO
      P(N,1)=ZERO
      DO 70 J=2,M
          J1=J-1
          P(1,J)=P(1,J1)+DY(J1)*V(1,J1)
          P(N,J)=P(N,J1)+DY(J1)*V(N1,J1)
70    CONTINUE
80    CALL CUB2D(N,M,NDIM,MDIM,X,Y,U,IR,EPS,P,Q,R,A,IFLAG,
     +           DX,AX,BX,CX,RX,DY,AY,BY,CY,RY)
      IF(IFLAG.NE.0) RETURN
      DO 120 I=1,N1
          DO 110 J=1,M1
              DO 100 K=1,3
                  DO 90 L=1,3
                      A(I,J,K,L)=K*L*A(I,J,K+1,L+1)
```

(*cont.*)
```
90                   CONTINUE
100              CONTINUE
110          CONTINUE
120 CONTINUE
    RETURN
    END
```

Figure 4.26. Program listing of QVASPL.

Calling sequence:

CALL QVASPL(N,M,NDIM,MDIM,X,Y,V,U,IR,EPS,P,Q,R,A,
            IFLAG,DX,AX,BX,CX,RX,DY,AY,BY,CY,RY)

Purpose:
For given points (x_i, y_j) $i = 1, \cdots, n \geq 2$; $j = 1, \cdots, m \geq 2$ in the plane with $x_1 < \cdots < x_n$ and $y_1 < \cdots < y_m$ and associated heights v_{ij}, this routine determines the coefficients $a_{ijk\ell}$ of a biquadratic histospline, F, i.e., F is of the form (3.1) and satisfies condition (4.49) on each of the subrectangles R_{ij}. For IR=1, most of the boundary values needed for this calculation are automatically set to the values of (4.54). However, values for R(1,1), R(N,1), R(1,M) and R(N,M) must be supplied by the user (one could, for example, use the values of (4.56)). For IR=2, all the required boundary values must be supplied by the user.

Description of the parameters:

N,M,NDIM,MDIM,X,Y,IR,EPS,IFLAG,
DX,AX,BX,CX,RX,DY,AY,BY,CY,RY as in CUB2D.
V   ARRAY(NDIM,M): Volume heights v_{ij}, $i = 1, \cdots, n - 1$;
           $j = 1, \cdots, m - 1$.
A   ARRAY(NDIM,MDIM,4,4): Output: Spline coefficients.
           A is first used, as dimensioned, as one of the arrays passed
           to CUB2D. Subsequently, the coefficients of the biquadratic
           histospline are stored in the subarray A(NDIM,MDIM,3,3).
U,P,Q,R   ARRAY(NDIM,M): Work space.

Required subroutines: CUB2D, TRDISA, CBPERM, TRDISB,
CUBMAT.

Figure 4.27. Description of QVASPL.

```
      FUNCTION QVAVAL(U,V,N,NDIM,M,MDIM,X,Y,A,IFLAG)
      REAL U,V,X(N),Y(M),A(NDIM,MDIM,4,4)
      DATA I,J/2*1/
      ZERO=0.
      IFLAG=0
      IF(N.LT.2.OR.M.LT.2) THEN
          IFLAG=1
          RETURN
      END IF
      CALL INTTWO(X,N,Y,M,U,V,I,J,IFLAG)
      IF(IFLAG.NE.0) RETURN
      UX=U-X(I)
      VY=V-Y(J)
      QVAVAL=ZERO
      DO 20 K=3,1,-1
          QV=A(I,J,K,3)
          DO 10 L=2,1,-1
              QV=QV*VY+A(I,J,K,L)
10        CONTINUE
          QVAVAL=QVAVAL*UX+QV
20    CONTINUE
      RETURN
      END
```

Figure 4.28. Program listing of QVAVAL.

This method is implemented in QVASPL (Figs. 4.26 and 4.27). The FUNCTION QUAVAL (Fig. 4.28), intended to be used for evaluation, is identical to QUMVAL except for the different dimensioning of A. This is required in QVASPL, since it calls CUB2D (see also the remark in the description of RVASPL). Hence, it does not require its own program description.

The data of the first three examples on histo-interpolation, shown in Figs. 4.29–4.31, originate from the earlier examples on bicubic spline interpolation, 4.7, 4.10, and 4.14. The x_i and y_j are the same and the v_{ij} were set to $v_{ij} = u_{ij}$, $i = 1, \cdots, n-1$; $j = 1, \cdots, m-1$. The figures have a somewhat different appearance due to the fact that u_{n1}, u_{nm}, and u_{1m} are not involved and because of the (different) boundary values (4.56). These latter were used for all the plots. Also, one should compare Fig. 3.7 with 4.30, and Fig. 3.9 with 4.31. In Fig. 4.32, the v_{ij} were set to

$$\begin{pmatrix} 2 & 3 & 4 & 5 & 6 \\ 3 & 3 & 4 & 5 & 6 \\ 3 & 4 & 3 & 6 & 7 \end{pmatrix},$$

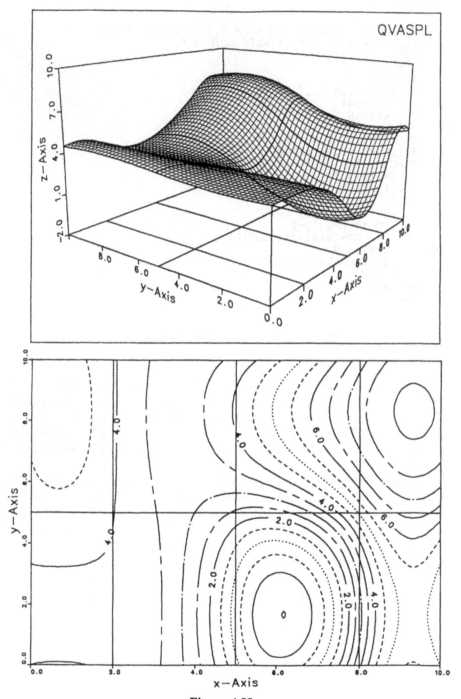

Figure 4.29.

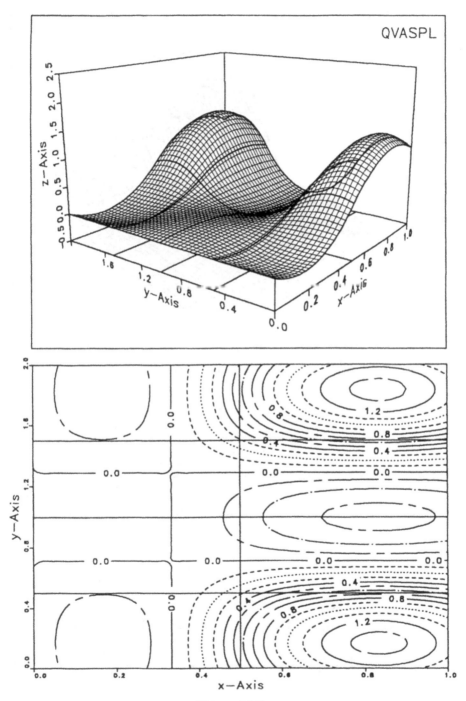

Figure 4.30.

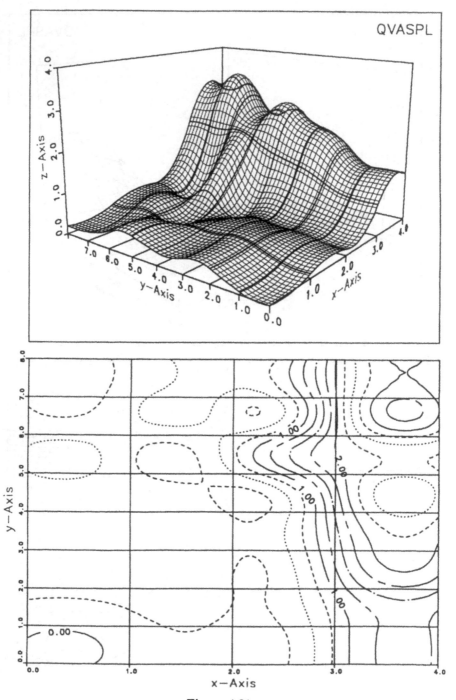

Figure 4.31.

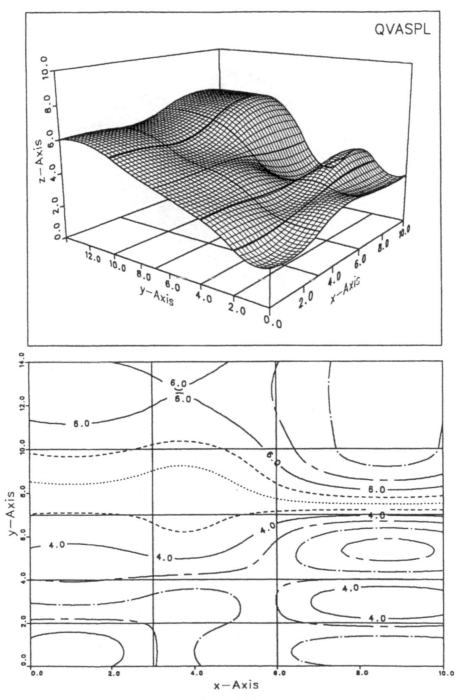

Figure 4.32.

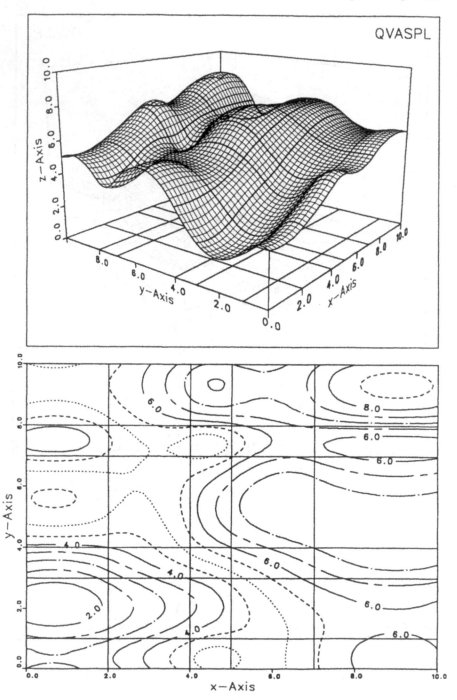

Figure 4.33.

and in Fig. 4.33 to

$$\begin{pmatrix} 3 & 2 & 3 & 5 & 4 & 5 \\ 4 & 3 & 4 & 5 & 5 & 6 \\ 5 & 4 & 5 & 6 & 5 & 7 \\ 5 & 5 & 6 & 7 & 6 & 7 \\ 6 & 6 & 7 & 7 & 6 & 8 \end{pmatrix}.$$

In both cases, the x_i and y_j can be read off the axes.

Other bivariate polynomial histosplines are treated in detail in [102], namely, those of degree four that then have continuous derivatives of the third order or alternatively only of the second order but with additional prescribed values to be interpolated at the grid points. This generalizes the univariate case as discussed in [101, p. 59 ff.] and [100].

5

Birational Spline Interpolants

5.1. Birational Spline Interpolants on Rectangular Grids

A number of variants of univariate spline interpolation using simple rational functions are considered in [100, Chapt. 6]. It seems unlikely that adaptive rational splines can be usefully generalized, or at least it would be rather difficult to do so seeing that, on each subrectangle R_{ij}, condition (6.44) of [100] might or might not hold in both x and y directions at the same time or it might hold in one direction but not in the other. This latter case is treated, however, in the research article ([130] in the bibliography of [100]). appeared in print. Rational C^2 spline interpolants with a single prescribable pole ([100], Chapt. 6) have not proven to be particularly useful. This is also true for for those with two given real or complex poles ([100], Chapt. 6). The generalization of shape-preserving rational spline interpolation ([100]) does not seem to be particularly straightforward. For the time being, what remains is the generalization of C^2 spline interpolation with two prescribable real poles ([100]). This has already been considered in [101].

We thus return to the form (4.1), except with

$$g_{i1}(x) \quad = \quad g_{i1}(x, x_i, x_{i+1}) = \frac{x_{i+1} - x}{\Delta x_i},$$

$$g_{i2}(x) \;=\; g_{i2}(x, x_i, x_{i+1}) = \frac{x - x_i}{\Delta x_i},$$

$$g_{i3}(x) \;=\; g_{i3}(x, x_i, x_{i+1}, px_i) = \frac{\left(\frac{x_{i+1}-x}{\Delta x_i}\right)^3}{1 + px_i \left(\frac{x-x_i}{\Delta x_i}\right)}, \qquad (5.1)$$

$$g_{i4}(x) \;=\; g_{i4}(x, x_i, x_{i+1}, qx_i) = \frac{\left(\frac{x-x_i}{\Delta x_i}\right)^3}{1 + qx_i \left(\frac{x_{i+1}-x}{\Delta x_i}\right)},$$

where

$$px_i, qx_i > -1, \quad i = 1, \cdots, n-1, \qquad (5.2)$$

are numbers that determine the positions of the poles. By the same
functions with argument y instead of x, we mean those where all the x-arguments have been replaced by the corresponding y-arguments, so that,
for example,

$$g_{j3}(y) = g_{j3}(y, y_j, y_{j+1}, py_j) = \frac{\left(\frac{y_{j+1}-y}{\Delta y_j}\right)^3}{1 + py_j \left(\frac{y-y_j}{\Delta y_j}\right)}.$$

These require the additional numbers,

$$py_j, qy_j > -1, \quad j = 1, \cdots, m-1. \qquad (5.3)$$

If $'$ means differentiation with respect to x (respectively, y), then

$$g'_{i1}(x) \;=\; -\frac{1}{\Delta x_i},$$

$$g'_{i2}(x) \;=\; \frac{1}{\Delta x_i}, \qquad (5.4)$$

$$g'_{i3}(x) \;=\; \frac{1}{\Delta x_i} \frac{2px_i \left(\frac{x_{i+1}-x}{\Delta x_i}\right)^3 - 3(1 + px_i)\left(\frac{x_{i+1}-x}{\Delta x_i}\right)^2}{\left(1 + px_i \left(\frac{x-x_i}{\Delta x_i}\right)\right)^2},$$

$$g'_{i4}(x) \;=\; -\frac{1}{\Delta x_i} \frac{2qx_i \left(\frac{x-x_i}{\Delta x_i}\right)^3 - 3(1 + qx_i)\left(\frac{x-x_i}{\Delta x_i}\right)^2}{\left(1 + qx_i \left(\frac{x_{i+1}-x}{\Delta x_i}\right)\right)^2}.$$

With these, we may compute the *connection matrix* of (4.5) to be

$$V(x_i) = V(x_i, x_{i+1}, px_i, qx_i) = \begin{pmatrix} 1 & 0 & 1 & 0 \\ -\frac{1}{\Delta x_i} & \frac{1}{\Delta x_i} & -\frac{3+px_i}{\Delta x_i} & 0 \\ 0 & 1 & 0 & 1 \\ -\frac{1}{\Delta x_i} & \frac{1}{\Delta x_i} & 0 & \frac{3+qx_i}{\Delta x_i} \end{pmatrix}.$$

$$(5.5)$$

Setting

$$\alpha = 3 + px_i,$$
$$\beta = 3 + qx_i, \tag{5.6}$$
$$\gamma = \alpha\beta - \alpha - \beta,$$

its inverse in general may be calculated ([101]) to be

$$[V(x_i)]^{-1} = \frac{1}{\gamma}\begin{pmatrix} \alpha(\beta-1) & (\beta-1)\Delta x_i & -\beta & \Delta x_i \\ -\alpha & -\Delta x_i & \beta(\alpha-1) & -(\alpha-1)\Delta x_i \\ -\beta & -(\beta-1)\Delta x_i & \beta & -\Delta x_i \\ \alpha & \Delta x_i & -\alpha & (\alpha-1)\Delta x_i \end{pmatrix}. \tag{5.7}$$

Naturally, in the corresponding matrix $[V(y_j)]^{-1}$, we must take $\alpha = 3+py_j$ and $\beta = 3 + qy_j$.

For the special case of

$$px_i = qx_i = p > -1, \quad i = 1, \cdots, n-1, \tag{5.8}$$

(5.7) simplifies to

$$[V(x_i)]^{-1} = \frac{1}{1+p}\begin{pmatrix} 2+p & \frac{2+p}{3+p}\Delta x_i & -1 & \frac{1}{3+p}\Delta x_i \\ -1 & -\frac{1}{3+p}\Delta x_i & 2+p & -\frac{2+p}{3+p}\Delta x_i \\ -1 & -\frac{2+p}{3+p}\Delta x_i & 1 & -\frac{1}{3+p}\Delta x_i \\ 1 & \frac{1}{3+p}\Delta x_i & -1 & \frac{2+p}{3+p}\Delta x_i \end{pmatrix}. \tag{5.9}$$

In both cases then, the coefficients of our birational spline function satisfy (4.8), where C is the matrix of (4.4) involving the parameters q_{ij}, p_{ij}, r_{ij}, $q_{i,j+1}$, $p_{i,j+1}$, $r_{i,j+1}$, $q_{i+1,j}$, $p_{i+1,j}$, $r_{i+1,j}$, $q_{i+1,j+1}$, $p_{i+1,j+1}$, and $r_{i+1,j+1}$.

As boundary conditions, we again assume that (4.11) and (4.12) are given. Using [100, formula (6.70)], the C^2 requirement in the x direction yields for each $j = 1, \cdots, m$, the system of equations,

$$\frac{qx_{i-1}^2 + 3qx_{i-1} + 3}{(2 + px_{i-1})(2 + qx_{i-1}) - 1}\frac{1}{\Delta x_{i-1}}p_{i-1,j} +$$
$$\left[\frac{(qx_{i-1}^2 + 3qx_{i-1} + 3)(2 + px_{i-1})}{(2 + px_{i-1})(2 + qx_{i-1}) - 1}\frac{1}{\Delta x_{i-1}}\right.$$
$$\left. + \frac{(px_i^2 + 3px_i + 3)(2 + qx_i)}{(2 + px_i)(2 + qx_i) - 1}\frac{1}{\Delta x_i}\right]p_{ij}$$
$$+ \frac{px_i^2 + 3px_i + 3}{(2 + px_i)(2 + qx_i) - 1}\frac{1}{\Delta x_i}p_{i+1,j} \tag{5.10}$$
$$= \frac{(qx_{i-1}^2 + 3qx_{i-1} + 3)(3 + px_{i-1})}{(2 + px_{i-1})(2 + qx_{i-1}) - 1}\frac{u_{ij} - u_{i-1,j}}{\Delta x_{i-1}^2}$$

$$+\frac{(px_i^2 + 3px_i + 3)(3 + qx_i)}{(2 + px_i)(2 + qx_i) - 1}\frac{u_{i+1,j} - u_{ij}}{\Delta x_i^2}, \quad i = 2, \cdots, n - 1.$$

These are analogous to (4.9). Similarly, in the y direction, we obtain for each $i = 1, \cdots, n$, the systems analogous to (4.10),

$$
\begin{aligned}
&\frac{qy_{j-1}^2 + 3qy_{j-1} + 3}{(2 + py_{j-1})(2 + qy_{j-1}) - 1}\frac{1}{\Delta y_{j-1}}q_{i,j-1} + \\
&\left[\frac{(qy_{j-1}^2 + 3qy_{j-1} + 3)(2 + py_{j-1})}{(2 + py_{j-1})(2 + qy_{j-1}) - 1}\frac{1}{\Delta y_{j-1}}\right. \\
&\quad \left. + \frac{(py_j^2 + 3py_j + 3)(2 + qy_j)}{(2 + py_j)(2 + qy_j) - 1}\frac{1}{\Delta y_j}\right]q_{ij} \\
&\quad + \frac{py_j^2 + 3py_j + 3}{(2 + py_j)(2 + qy_j) - 1}\frac{1}{\Delta y_j}q_{i,j+1} \\
&= \frac{(qy_{j-1}^2 + 3qy_{j-1} + 3)(3 + py_{j-1})}{(2 + py_{j-1})(2 + qy_{j-1}) - 1}\frac{u_{ij} - u_{i,j-1}}{\Delta y_{j-1}^2} \\
&\quad + \frac{(py_j^2 + 3py_j + 3)(3 + qy_j)}{(2 + py_j)(2 + qy_j) - 1}\frac{u_{i,j+1} - u_{ij}}{\Delta y_j^2}, \quad j = 2, \cdots, m - 1.
\end{aligned}
\tag{5.11}
$$

We will not explicitly write out the systems for the r_{ij} corresponding to (4.13) and (4.14). These are obtained by simply replacing p by r and u by q for $j = 1$ and $j = m$ in (5.10) and replacing q by r and u by p for $i = 1, \cdots, n$ in (5.11). Again, there are altogether only two different coefficient matrices involved. They are both tridiagonal and strictly diagonally dominant but in general not symmetric. Hence, existence and uniqueness of the solution is guaranteed. In the special case of

$$px_i = qx_i = py_j = qy_j = p > -1, \tag{5.12}$$

considered in [101], the systems simplify dramatically. For example (5.10) becomes

$$\frac{1}{\Delta x_{i-1}}p_{i-1,j} + (2 + p)\left[\frac{1}{\Delta x_{i-1}} + \frac{1}{\Delta x_i}\right]p_{ij} + \frac{1}{\Delta x_i}p_{i+1,j}$$

$$= (3 + p)\left[\frac{u_{ij} - u_{i-1,j}}{\Delta x_{i-1}^2} + \frac{u_{i+1,j} - u_{ij}}{\Delta x_i^2}\right], \quad i = 2, \cdots, n - 1. \tag{5.13}$$

The coefficient matrix is now symmetric. For the special case of $p = 0$, we recover (4.9). Consequently bicubic spline interpolation is a special case of birational interpolation, with the only difference being that the basis functions are different from those of (3.2) and (4.2). Indeed, they are sufficiently different so that we can no longer use a bivariate Horner's rule.

That such birational spline interpolants are always (i.e., not only for $p = 0$) C^2 on all of R, i.e., also across interior gridlines, can be shown in the same way that this was shown for bicubic splines, just following formula (4.14). By construction, the function $\partial^4 F/\partial x^2 \partial y^2$ is again continuous (but not bilinear) on all of R. On the sides of R_{ij}, it restricts to a function of the form, $b_3 g''_{i3} + b_4 g''_{i4}$, which, similar to linears, has the property that the two coefficients, b_3 and b_4, are uniquely determined by given function values at two different points. Further, we must use use the fact that its second integral can be written in the form, $a_1 g_{i1} + a_2 g_{i2} + a_3 g_{i3} + a_4 g_{i4}$, and that these four coefficients $a_1, \cdots, a_4$ are uniquely determined by function values and first derivative values at two different points (see [100, formula (6.69)].

From the results of [100], it is also clear that as px_i, qx_i, py_j, and $qy_j \to \infty$ simultaneously, the C^2 birational spline interpolant tends to a bilinear function. This effect can be used, just as in the univariate case, for local shape preservation. This will be demonstrated in the examples.

We now first describe a subroutine RAT2D for the special case (5.12) and then one called RSP2D for the general case (5.2) and (5.3). RAT2D (Figs. 5.1 and 5.2) has exactly the same structure as CUB2D, and RT-PERM (Fig. 5.3) and RATMAT (Fig. 5.4) have the same functions as CBPERM and CUBMAT, respectively.

The FUNCTION RBIVAL (Figs. 5.5 and 5.6), like CBIVAL, is for evaluation of C^2 rational spline interpolants for the case of (5.12). It, however, uses an additional FUNCTION GI (Fig. 5.7) that calculates the functions (5.1).

The structure of RSP2D (Figs. 5.8 and 5.9) is somewhat more complicated. The required coefficient matrices are assembled by means of RSPDEC (Fig. 5.10). TRDIUA (see the appendix) computes the LU-decomposition of the now nonsymmetric tridiagonal coefficient matrices. RSPPER (Fig. 5.11) forms the right-hand sides and calculates the corresponding solutions using the precomputed LU-decompositions as well as TRDIUB (see the appendix). RSPMAT (Fig. 5.12) sets the matrices (5.7), and finally, RSPVAL (Figs. 5.13 and 5.14) may be used for evaluation. It calls the FUNCTION GIJ (Fig. 5.15).

Figures 5.16–5.24 were produced with RAT2D. They correspond to Figs. 4.6, 4.8, 4.9, 4.11, and 4.13–4.17. The values of p that were used are shown on the figures themselves. For $p = 10$, Fig. 5.18 seems already rather bilinear, and Fig. 5.24 is less wavy than 4.16. The bimonotonicity of the data is obviously not yet reproduced in Fig. 5.22. Hence we tried, in RSP2D, to find suitable, locally dependent values for the px_i, qx_i, py_j, and qy_j. The last example, Fig. 5.25, corresponds to

px_i	qx_i	py_j	qy_j
10	10	50	50
50	50	750	750
750	750	750	750
750	750	50	50
50	50	10	10
10	10	10	10
10	10		
10	10		
10	10		
10	10		

The interpolant now appears to be bimonotone (indeed almost bilinear), but a close inspection of the numerical values reveals that this is not entirely so. This could also be because of rounding errors in the calculation caused by the large values of the px_i, qx_i, py_j, and qy_j. Even for less problematic test examples, it is not entirely easy to find suitable tension parameters. For values too large (roughly starting at 10), there is always the tendency for the surface to appear too bilinear. Nevertheless, one can, with RSP2D, attempt to have shape preservation by locally adjusting the parameters.

Just as in the case of bicubic splines, it might be better to use *birational* C^1 *Hermite spline interpolation*. Once it is understood how to go from CUB2D to HCUB2D, it is very easy to write subroutines that correspond to RAT2D and RSP2D. Then both local tension parameters *and* values for the derivatives can be used to control the shape of the surfaces.

As we saw in the univariate case ([100, Chapt. 7]), exponential splines can be obtained in a manner very similar to that for rational splines except that they are, computationally, substantially more expensive. Hence, we will not give bivariate subroutines for these. Their theory may be found in [101, pp. 113–114], and an ALGOL procedure is given in [99]. There are further discussions of this in [75].

5.2. Birational Histosplines

At the end of the previous chapter, we were very easily able to obtain C^1 biquadratic histosplines from C^2 bicubic spline interpolants. Just as in the univariate case ([100]), this kind of derivation is not limited to the use of polynomials ([102]). For example, we could choose

```
      SUBROUTINE RAT2D(N,M,NDIM,MDIM,X,Y,U,IR,PP,EPS,P,Q,R,A,
     +               IFLAG,DX,AX,BX,CX,RX,DY,AY,BY,CY,RY)
      DIMENSION X(2),Y(2),U(NDIM,M),P(NDIM,M),Q(NDIM,M),
     +          R(NDIM,M),A(NDIM,MDIM,4,4),
     +          DX(N),AX(N),BX(N),CX(N),RX(N),
     +          DY(M),AY(M),BY(M),CY(M),RY(M),
     +          B(4,4),C(4,4),D(4,4),E(4,4)
      IFLAG=0
      IF(N.LT.2.OR.M.LT.2) THEN
          IFLAG=1
          RETURN
      END IF
      IF(PP.LT.(EPS-1.)) THEN
          IFLAG=3
          RETURN
      END IF
      N1=N-1
      N2=N-2
      M1=M-1
      M2=M-2
      ZERO=0.
      DO 10 I=1,N1
          DX(I)=1./(X(I+1)-X(I))
10    CONTINUE
      DO 20 J=1,M1
          DY(J)=1./(Y(J+1)-Y(J))
20    CONTINUE
      P1=PP+1.
      P2=PP+2.
      P3=PP+3.
      GA=P3*P1
      DK=1/P1
      B(1,1)=P2*DK
      B(2,1)=-DK
      B(4,3)=-DK
      B(4,1)=DK
      B(2,3)=B(1,1)
      B(1,3)=-DK
      B(3,1)=-DK
      B(3,3)=DK
      DO 40 J=1,3,2
          DO 30 I=1,4
              E(I,J)=B(I,J)
30        CONTINUE
40    CONTINUE
      IF(IR.EQ.2) GOTO 70
      DO 50 J=1,M
          P(1,J)=(U(2,J)-U(1,J))*DX(1)
          P(N,J)=(U(N,J)-U(N1,J))*DX(N1)
50    CONTINUE
      DO 60 I=1,N
          Q(I,1)=(U(I,2)-U(I,1))*DY(1)
          Q(I,M)=(U(I,M)-U(I,M1))*DY(M1)
60    CONTINUE
      R(1,1)=((P(1,2)-P(1,1))*DY(1)+(Q(2,1)-Q(1,1))*DX(1))/2.
      R(1,M)=((P(1,M)-P(1,M1))*DY(M1)+(Q(2,M)-Q(1,M))*DX(1))/2.
      R(N,1)=((P(N,2)-P(N,1))*DY(1)+(Q(N,1)-Q(N1,1))*DX(N1))/2.
```

(cont.)

```
      R(N,M)=((P(N,M)-P(N,M1))*DY(M1)+(Q(N,M)-Q(N1,M))*DX(N1))/2.
70    IF(N.EQ.2.AND.M.EQ.2) GOTO 130
      IF(N.EQ.2) GOTO 90
      DO 80 I=1,N2
         IF(I.LT.N2) AX(I)=DX(I+1)
         BX(I)=P2*(DX(I+1)+DX(I))
80    CONTINUE
      CALL TRDISA(N2,AX,BX,CX,EPS,IFLAG)
      IF(IFLAG.NE.0) RETURN
      IVJ=0
      CALL RTPERM(IVJ,M,1,N,NDIM,N1,N2,P,U,P3,AX,BX,CX,DX,RX)
90    IF(M.EQ.2) GOTO 110
      DO 100 J=1,M2
         IF(J.LT.M2) AY(J)=DY(J+1)
         BY(J)=P2*(DY(J+1)+DY(J))
100   CONTINUE
      CALL TRDISA(M2,AY,BY,CY,EPS,IFLAG)
      IF(IFLAG.NE.0) RETURN
      IVJ=1
      CALL RTPERM(IVJ,N,1,M,NDIM,M1,M2,Q,U,P3,AY,BY,CY,DY,RY)
110   IF(N.EQ.2) GOTO 120
      IVJ=0
      CALL RTPERM(IVJ,M,M1,N,NDIM,N1,N2,R,Q,P3,AX,BX,CX,DX,RX)
120   IF(M.EQ.2) GOTO 130
      IVJ=1
      CALL RTPERM(IVJ,N,1,M,NDIM,M1,M2,R,P,P3,AY,BY,CY,DY,RY)
130   DO 210 I=1,N1
         I1=I+1
         CALL RATMAT(1./DX(I),PP,B)
         DO 200 J=1,M1
            J1=J+1
            C(1,1)=U(I,J)
            C(1,2)=Q(I,J)
            C(2,1)=P(I,J)
            C(2,2)=R(I,J)
            C(1,3)=U(I,J1)
            C(1,4)=Q(I,J1)
            C(2,3)=P(I,J1)
            C(2,4)=R(I,J1)
            C(3,1)=U(I1,J)
            C(3,2)=Q(I1,J)
            C(4,1)=P(I1,J)
            C(4,2)=R(I1,J)
            C(3,3)=U(I1,J1)
            C(3,4)=Q(I1,J1)
            C(4,3)=P(I1,J1)
            C(4,4)=R(I1,J1)
            DO 160 K1=1,4
               DO 150 K2=1,4
                  SUM=ZERO
                  DO 140 K=1,4
                     SUM=SUM+B(K1,K)*C(K,K2)
140               CONTINUE
                  D(K1,K2)=SUM
150            CONTINUE
160         CONTINUE
            CALL RATMAT(1./DY(J),PP,E)
```

(*cont.*)

```
            DO 190 K1=1,4
                DO 180 K2=1,4
                    SUM=ZERO
                    DO 170 K=1,4
                        SUM=SUM+D(K1,K)*E(K2,K)
170                 CONTINUE
                    A(I,J,K1,K2)=SUM
180             CONTINUE
190         CONTINUE
200     CONTINUE
210 CONTINUE
    RETURN
    END
```

Figure 5.1. Program listing of RAT2D.

Calling sequence:

CALL RAT2D(N,M,NDIM,MDIM,X,Y,U,IR,PP,EPS,P,Q,R,A,
    IFLAG,DX,AX,BX,CX,RX,DY,AY,BY,CY,RY)

Purpose:

For given points (x_i, y_j) $i = 1, \cdots, n \geq 2$; $j = 1, \cdots, m \geq 2$ in the plane with $x_1 < \cdots < x_n$ and $y_1 < \cdots < y_m$ and associated heights u_{ij}, this routine determines the coefficients $a_{ijk\ell}$ of a special birational spline interpolant F, which on each subrectangle R_{ij} is of the form (4.1) with g_{ij} given by (5.1), $i = 1, \cdots, n - 1$; $j = 1, \cdots, 4$. The parameters are set according to (5.12). For IR=1, the boundary values needed for this calculation are automatically computed by means of simple difference quotients. For IR=2, the boundary values for the partial derivatives with respect to x, p_{ij}, $i = 1, n$; $j = 1, \cdots, m$, and with respect to y, q_{ij}, $i = 1, \cdots, n$; $j = 1, m$, as well as the values r_{11}, r_{1m}, r_{n1}, and r_{nm} for the mixed partials, must be supplied by the user.

Description of the parameters:

N,M,NDIM,MDIM,X,Y,U,IR,EPS,P,Q,R,A,
DX,AX,BX,CX,RX,DY,AY,BY,CY,RY as in CUB2D.

PP	ARRAY(NDIM,M): Input: The parameters $p > -1$ according to (5.12).	
IFLAG	=0:	Normal execution.
	=1:	N<2 or M<2.
	=2:	Error in solving the linear system (TRDISA).
	=3:	$p \leq -1$.

Required subroutines: TRDISA, RTPERM, TRDISB, RATMAT.

Figure 5.2. Description of RAT2D.

```
      SUBROUTINE RTPERM(IVJ,M,M1,N,NDIM,N1,N2,P,U,P3,AX,BX,
     +               CX,DX,RX)
      DIMENSION P(NDIM,M),U(NDIM,M),AX(N),BX(N),CX(N),DX(N),
     +          RX(N)
      DO 40 J=1,M,M1
          DO 20 I=1,N1
              I1=I-1
              IF(IVJ.NE.0) THEN
                  R2=P3*DX(I)*DX(I)*(U(J,I+1)-U(J,I))
                  IF(I.EQ.1) GOTO 10
                  RX(I1)=R1+R2
                  IF(I.EQ.2) RX(I1)=RX(I1)-DX(1)*P(J,1)
                  IF(I.EQ.N1) RX(I1)=RX(I1)-DX(N1)*P(J,N)
              ELSE
                  R2=P3*DX(I)*DX(I)*(U(I+1,J)-U(I,J))
                  IF(I.EQ.1) GOTO 10
                  RX(I1)=R1+R2
                  IF(I.EQ.2) RX(I1)=RX(I1)-DX(1)*P(1,J)
                  IF(I.EQ.N1) RX(I1)=RX(I1)-DX(N1)*P(N,J)
              END IF
10            R1=R2
20        CONTINUE
          CALL TRDISB(N2,AX,BX,CX,RX)
          DO 30 I=2,N1
              IF(IVJ.EQ.0) P(I,J)=RX(I-1)
              IF(IVJ.NE.0) P(J,I)=RX(I-1)
30        CONTINUE
40    CONTINUE
      RETURN
      END
```

Figure 5.3. Program listing of RTPERM.

```
      SUBROUTINE RATMAT(H,PP,B)
      DIMENSION B(4,4)
      FA=H/(PP+3.)
      B(1,2)=B(1,1)*FA
      B(2,2)=B(2,1)*FA
      B(3,2)=-B(1,2)
      B(4,2)=-B(2,2)
      B(1,4)=B(4,2)
      B(2,4)=B(2,2)*(PP+2.)
      B(3,4)=B(2,2)
      B(4,4)=-B(2,4)
      RETURN
      END
```

Figure 5.4. Program listing of RATMAT.

```
      FUNCTION RBIVAL(U,V,N,NDIM,M,MDIM,X,Y,A,PP,IFLAG)
      DIMENSION X(N),Y(M),A(NDIM,MDIM,4,4)
      DATA I,J/2*1/
      IF(N.LT.2.OR.M.LT.2) THEN
          IFLAG=1
          RETURN
      END IF
      CALL INTTWO(X,N,Y,M,U,V,I,J,IFLAG)
      IF(IFLAG.NE.0) RETURN
      UX=(U-X(I))/(X(I+1)-X(I))
      VY=(V-Y(J))/(Y(J+1)-Y(J))
      RBIVAL=0.
      DO 20 K=1,4
          H=GI(K,UX,PP)
          DO 10 L=1,4
              RBIVAL=RBIVAL+A(I,J,K,L)*H*GI(L,VY,PP)
10        CONTINUE
20    CONTINUE
      RETURN
      END
```

Figure 5.5. Program listing of RBIVAL.

FUNCTION RBIVAL(V,W,N,NDIM,M,MDIM,X,Y,A,PP,IFLAG)

Purpose:
Calculation of a function value of a special (see description of RAT2D) birational spline interpolant at a point $(v, w) \in R$, where R is the underlying rectangular grid.

Description of the parameters:

V,W,N,NDIM,M,MDIM,X,Y,A,IFLAG as in CBIVAL.
PP   The parameter p (see RAT2D).

Required subroutines: INTTWO, GI.

Figure 5.6. Description of RBIVAL.

```
      FUNCTION GI(I,X,PP)
      H=1.-X
      IF(I.EQ.1) GI=H
      IF(I.EQ.2) GI=X
      IF(I.EQ.3) GI=H*H*H/(PP*X+1.)
      IF(I.EQ.4) GI=X*X*X/(PP*H+1.)
      RETURN
      END
```

Figure 5.7. Program listing of GI.

```
      SUBROUTINE RSP2D(N,M,NDIM,MDIM,X,Y,U,PX,QX,PY,QY,EPS1,
     +                 EPS2,IR,P,Q,R,A,IFLAG,DX,AX,BX,CX,RX,DY,
     +                 AY,BY,CY,RY)
      DIMENSION X(N),Y(M),U(NDIM,M),PX(N),QX(N),PY(M),QY(M),
     +          P(NDIM,M),Q(NDIM,M),R(NDIM,M),A(NDIM,MDIM,4,4),
     +          DX(N),AX(N),BX(N),CX(N),RX(N),
     +          DY(M),AY(M),BY(M),CY(M),RY(M),
     +          B(4,4),C(4,4),D(4,4),E(4,4)
      IFLAG=0
      IF(N.LT.2.OR.M.LT.2) THEN
          IFLAG=1
          RETURN
      END IF
      N1=N-1
      N2=N-2
      M1=M-1
      M2=M-2
      E1=EPS1-1.
      ZERO=0.
      DO 10 I=1,N1
          DX(I)=1./(X(I+1)-X(I))
 10   CONTINUE
      DO 20 J=1,M1
          DY(J)=1./(Y(J+1)-Y(J))
 20   CONTINUE
      IF(IR.EQ.2) GOTO 70
      DO 50 J=1,M
          P(1,J)=(U(2,J)-U(1,J))*DX(1)
          P(N,J)=(U(N,J)-U(N1,J))*DX(N1)
 50   CONTINUE
      DO 60 I=1,N
          Q(I,1)=(U(I,2)-U(I,1))*DY(1)
          Q(I,M)=(U(I,M)-U(I,M1))*DY(M1)
 60   CONTINUE
      R(1,1)=((P(1,2)-P(1,1))*DY(1)+(Q(2,1)-Q(1,1))*DX(1))/2.
      R(1,M)=((P(1,M)-P(1,M1))*DY(M1)+(Q(2,M)-Q(1,M))*DX(1))/2.
      R(N,1)=((P(N,2)-P(N,1))*DY(1)+(Q(N,1)-Q(N1,1))*DX(N1))/2.
      R(N,M)=((P(N,M)-P(N,M1))*DY(M1)+(Q(N,M)-Q(N1,M))*DX(N1))/2.
 70   IF(N.EQ.2.AND.M.EQ.2) GOTO 110
      IF(N.EQ.2) GOTO 80
      CALL RSPDEC(N1,PX,QX,DX,E1,AX,BX,CX,IFLAG)
      IF(IFLAG.NE.0) RETURN
      CALL TRDIUA(N2,AX,BX,CX,EPS2,IFLAG)
      IF(IFLAG.NE.0) RETURN
      IVJ=0
      CALL RSPPER(IVJ,M,1,N,NDIM,N1,N2,P,U,PX,QX,AX,BX,CX,DX,RX)
 80   IF(M.EQ.2) GOTO 90
      CALL RSPDEC(M1,PY,QY,DY,E1,AY,BY,CY,IFLAG)
      IF(IFLAG.NE.0) RETURN
      CALL TRDIUA(M2,AY,BY,CY,EPS2,IFLAG)
      IF(IFLAG.NE.0) RETURN
      IVJ=1
      CALL RSPPER(IVJ,N,1,M,NDIM,M1,M2,Q,U,PY,QY,AY,BY,CY,DY,RY)
```

(cont.)

```
90    IF(N.EQ.2) GOTO 100
      IVJ=0
      CALL RSPPER(IVJ,M,M1,N,NDIM,N1,N2,R,Q,PX,QX,AX,BX,CX,DX,RX)
100   IF(M.EQ.2) GOTO 110
      IVJ=1
      CALL RSPPER(IVJ,N,1,M,NDIM,M1,M2,R,P,PY,QY,AY,BY,CY,DY,RY)
110   DO 190 I=1,N1
          I1=I+1
          CALL RSPMAT(PX(I),QX(I),1./DX(I),B)
          DO 180 J=1,M1
              J1=J+1
              C(1,1)=U(I,J)
              C(1,2)=Q(I,J)
              C(2,1)=P(I,J)
              C(2,2)=R(I,J)
              C(1,3)=U(I,J1)
              C(1,4)=Q(I,J1)
              C(2,3)=P(I,J1)
              C(2,4)=R(I,J1)
              C(3,1)=U(I1,J)
              C(3,2)=Q(I1,J)
              C(4,1)=P(I1,J)
              C(4,2)=R(I1,J)
              C(3,3)=U(I1,J1)
              C(3,4)=Q(I1,J1)
              C(4,3)=P(I1,J1)
              C(4,4)=R(I1,J1)
              DO 140 K1=1,4
                  DO 130 K2=1,4
                      SUM=ZERO
                      DO 120 K=1,4
                          SUM=SUM+B(K1,K)*C(K,K2)
120                   CONTINUE
                      D(K1,K2)=SUM
130               CONTINUE
140           CONTINUE
              CALL RSPMAT(PY(J),QY(J),1./DY(J),E)
              DO 170 K1=1,4
                  DO 160 K2=1,4
                      SUM=ZERO
                      DO 150 K=1,4
                          SUM=SUM+D(K1,K)*E(K2,K)
150                   CONTINUE
                      A(I,J,K1,K2)=SUM
160               CONTINUE
170           CONTINUE
180       CONTINUE
190   CONTINUE
      RETURN
      END
```

Figure 5.8. Program listing of RSP2D.

Calling sequence:

CALL RSP2D(N,M,NDIM,MDIM,X,Y,U,PX,QX,PY,QY,EPS1,
           EPS2,IR,P,Q,R,A,IFLAG,DX,AX,BX,CX,RX,DY,
           AY,BY,CY,RY)

Purpose:

For given points (x_i, y_j) $i = 1, \cdots, n \geq 2$; $j = 1, \cdots, m \geq 2$ in the plane with $x_1 < \cdots < x_n$ and $y_1 < \cdots < y_m$ and associated heights u_{ij}, this routine determines the coefficients $a_{ijk\ell}$ of a general birational spline interpolant F, which on each subrectangle R_{ij} is of the form (4.1) with g_{ij} given by (5.1), $i = 1, \cdots, n-1$; $j = 1, \cdots, 4$. For IR=1, the boundary values needed for this calculation are automatically computed by means of simple difference quotients. For IR=2, the boundary values for the partial derivatives with respect to x, p_{ij}, $i = 1, n$; $j = 1, \cdots, m$, and with respect to y, q_{ij}, $i = 1, \cdots, n$; $j = 1, m$, as well as the values r_{11}, r_{1m}, r_{n1}, and r_{nm} for the mixed partials, must be supplied by the user.

Description of the parameters:

N,M,NDIM,MDIM,X,Y,U,IR,P,Q,R,A,
DX,AX,BX,CX,RX,DY,AY,BY,CY,RY as in RAT2D.

PX,QX	ARRAY(N): Parameters $px_i, qx_i > -1$. $i = 1, \cdots, n-1$.
PY,QY	ARRAY(N): Parameters $py_j, qy_j > -1$, $j = 1, \cdots, m-1$.
EPS1	Value used for accuracy test. The quantity $-1 +$ EPS1 is used as the parameter passed to RSPDEC. Inequalities (5.2) and (5.3) are to be satisfied, with $-1 +$ EPS1 replacing -1 on the right-hand sides. Recommendation: $0.1 \leq$ EPS1 ≤ 0.5.
EPS2	Value used for accuracy test. EPS2 is used as the parameter passed to TRDIUA.
IFLAG	=0: Normal execution.
	=1: N<2 or M<2.
	=2: Error in solving the linear system (TRDIUA).
	=3: $px_i \leq -1$, $qx_i \leq -1$, $py_j \leq -1$, or $qy_j \leq -1$ for some index.

Required subroutines: RSPDEC, TRDIUA, RSPPER, TRDIUB, RSPMAT.

Figure 5.9. Description of RSP2D.

```
      SUBROUTINE RSPDEC(N,PX,QX,DX,E1,AX,BX,CX,IFLAG)
      DIMENSION PX(N),QX(N),DX(N),AX(N),BX(N),CX(N)
      IFLAG=0
      DO 20 I=1,N
         I1=I-1
         PI=PX(I)
         QI=QX(I)
         IF(PI.LT.E1.OR.QI.LT.E1) THEN
            IFLAG=3
            RETURN
         END IF
         PI2=PI*(PI+3.)+3.
         QI2=QI*(QI+3.)+3.
         P22=2.+PI
         Q22=2.+QI
         G2=DX(I)/(P22*Q22-1.)
         IF(I.EQ.1) GOTO 10
         AX(I1)=QI2*G2
         BX(I1)=QI1*P21*G1+PI2*Q22*G2
         CX(I1)=PI2*G2
10       P21=P22
         QI1=QI2
         G1=G2
20    CONTINUE
      RETURN
      END
```

Figure 5.10. Program listing of RSPDEC.

```
      SUBROUTINE RSPPER(IVJ,M,M1,N,NDIM,N1,N2,P,U,PP,QQ,AX,BX,
     +                  CX,DX,RX)
      DIMENSION P(NDIM,M),U(NDIM,M),PP(N),QQ(N),
     +          AX(N),BX(N),CX(N),DX(N),RX(N)
      DO 40 J=1,M,M1
         DO 20 I=1,N1
            IF(IVJ.NE.0) THEN
               H=U(J,I+1)-U(J,I)
               IF(I.EQ.2) P1=P(J,1)
               IF(I.EQ.N1) P2=P(J,N)
            ELSE
               H=U(I+1,J)-U(I,J)
               IF(I.EQ.2) P1=P(1,J)
               IF(I.EQ.N1) P2=P(N,J)
            END IF
            I1=I-1
            PI=PP(I)
            QI=QQ(I)
            PI2=PI*(PI+3.)+3.
            QI2=QI*(QI+3.)+3.
            P22=2.+PI
            Q22=2.+QI
```

(cont.)

```
        G2=DX(I)/(P22*Q22-1.)
        R2=DX(I)*G2*H
        IF(I.EQ.1) GOTO 10
        RX(I1)=R1*QI1*(1.+P21)+R2*PI2*(1.+Q22)
        IF(I.EQ.2) RX(I1)=RX(I1)-QI1*G1*P1
        IF(I.EQ.N1) RX(I1)=RX(I1)-PI2*G2*P2
10      P21=P22
        QI1=QI2
        G1=G2
        R1=R2
20    CONTINUE
        CALL TRDIUB(N2,AX,BX,CX,RX)
        DO 30 I=2,N1
            IF(IVJ.EQ.0) P(I,J)=RX(I-1)
            IF(IVJ.NE.0) P(J,I)=RX(I-1)
30    CONTINUE
40  CONTINUE
    RETURN
    END
```

Figure 5.11. Program listing of RSPPER.

```
SUBROUTINE RSPMAT(PP,QQ,H,B)
DIMENSION B(4,4)
P2=PP+2.
Q2=QQ+2.
P3=PP+3.
Q3=QQ+3.
R=P3*Q3-P3-Q3
H1=Q2/R
H2=P2/R
B(1,1)=P3*H1
B(1,2)=H*H1
B(1,3)=-Q3/R
B(1,4)=H/R
B(2,1)=-P3/R
B(2,2)=-B(1,4)
B(2,3)=Q3*H2
B(2,4)=-H*H2
B(3,1)=B(1,3)
B(3,2)=-B(1,2)
B(3,3)=-B(1,3)
B(3,4)=B(2,2)
B(4,1)=-B(2,1)
B(4,2)=B(1,4)
B(4,3)=-B(4,1)
B(4,4)=-B(2,4)
RETURN
END
```

Figure 5.12. Program listing of RSPMAT.

```
FUNCTION RSPVAL(U,V,N,NDIM,M,MDIM,X,Y,A,PX,QX,PY,QY,IFLAG)
    DIMENSION X(N),Y(M),A(NDIM,MDIM,4,4),PX(N),QX(N),
   +          PY(M),QY(M)
    DATA I,J/2*1/
    IF(N.LT.2.OR.M.LT.2) THEN
        IFLAG=1
        RETURN
    END IF
    CALL INTTWO(X,N,Y,M,U,V,I,J,IFLAG)
    IF(IFLAG.NE.0) RETURN
    UX=(U-X(I))/(X(I+1)-X(I))
    VY=(V-Y(J))/(Y(J+1)-Y(J))
    PXI=PX(I)
    QXI=QX(I)
    PYJ=PY(J)
    QYJ=QY(J)
    RSPVAL=0.
    DO 20 K=1,4
        H=GIJ(K,UX,PXI,QXI)
        DO 10 L=1,4
            RSPVAL=RSPVAL+A(I,J,K,L)*H*GIJ(L,VY,PYJ,QYJ)
10      CONTINUE
20  CONTINUE
    RETURN
    END
```

Figure 5.13. Program listing of RSPVAL.

FUNCTION RSPVAL(V,W,N,NDIM,M,MDIM,X,Y,A,
                    PX,QX,PY,QY,IFLAG)

Purpose:
Calculation of a function value of a general (see description of RSP2D) birational spline interpolant at a point $(v, w) \in R$, where R is the underlying rectangular grid.

Description of the parameters:

V,W,N,NDIM,M,MDIM,X,Y,A,IFLAG as in CBIVAL.
PX,QX,PY,QY as in RSP2D.

Required subroutines: INTTWO, GIJ.

Figure 5.14. Description of RSPVAL.

```
FUNCTION GIJ(I,X,PP,QQ)
H=1.-X
IF(I.EQ.1) GIJ=H
IF(I.EQ.2) GIJ=X
IF(I.EQ.3) GIJ=H*H*H/(PP*X+1.)
IF(I.EQ.4) GIJ=X*X*X/(QQ*H+1.)
RETURN
END
```

Figure 5.15. Program listing of GIJ.

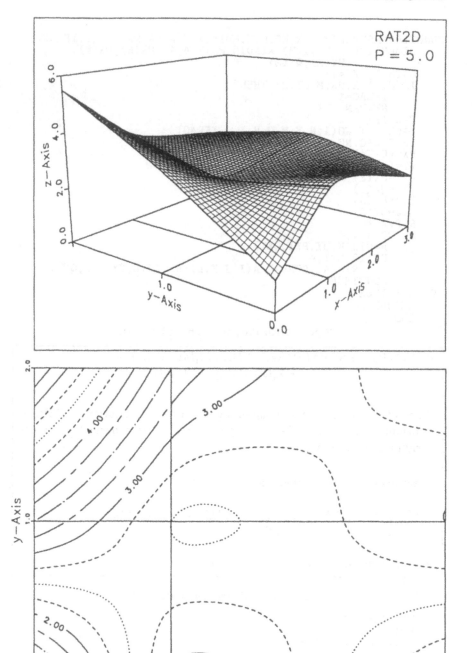

Figure 5.16.

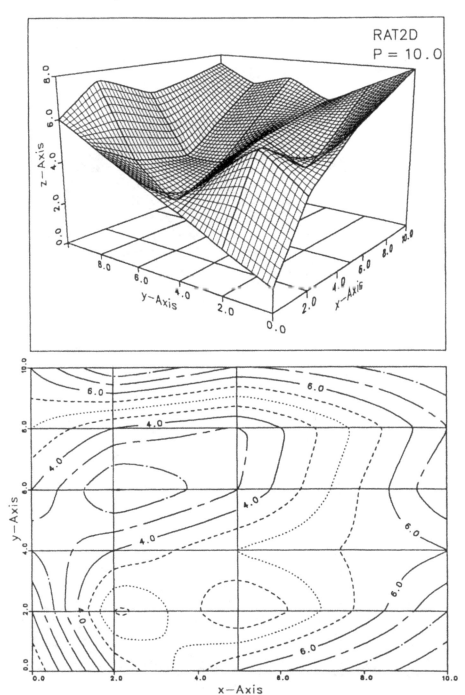

Figure 5.17.

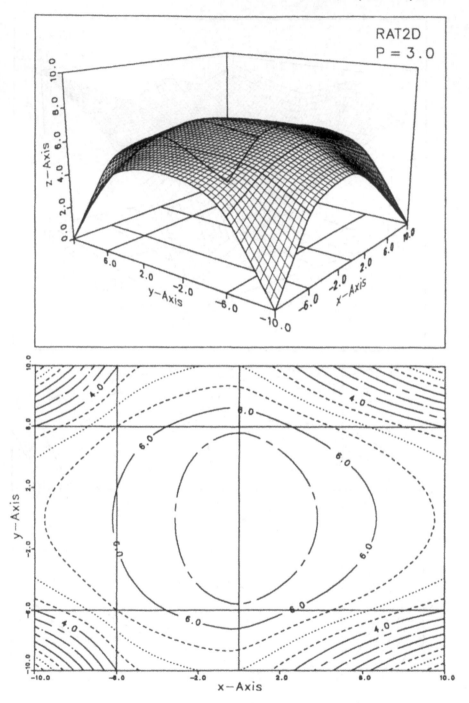

Figure 5.18.

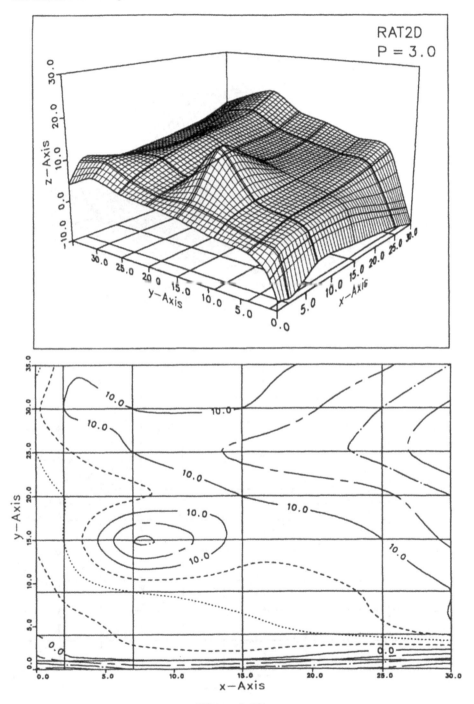

Figure 5.19.

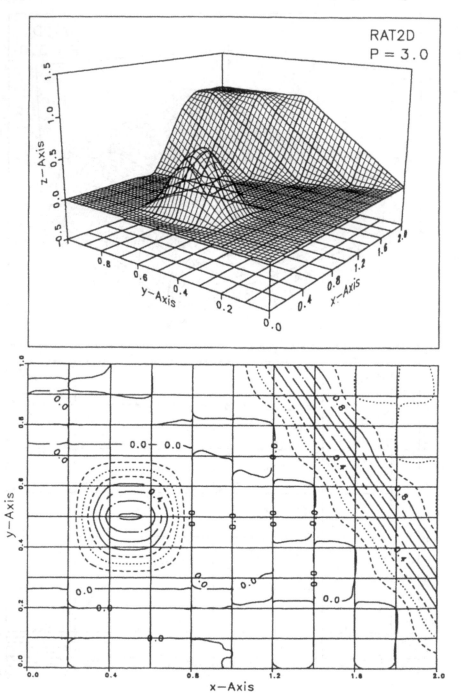

Figure 5.20.

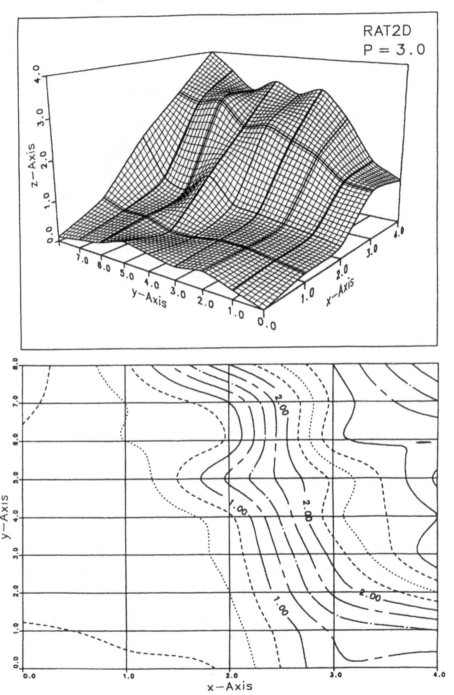

Figure 5.21.

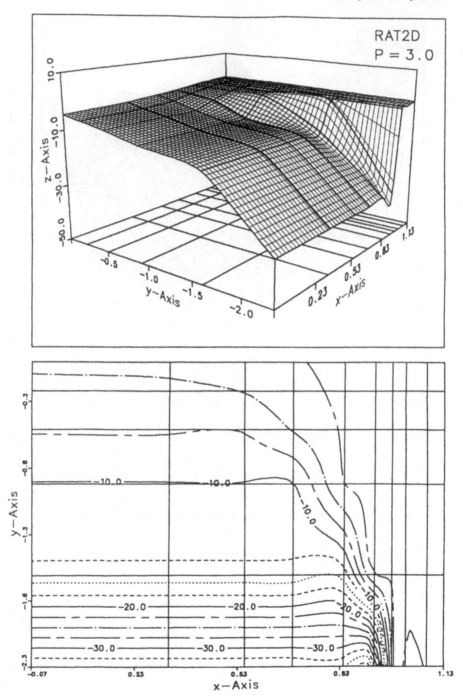

Figure 5.22.

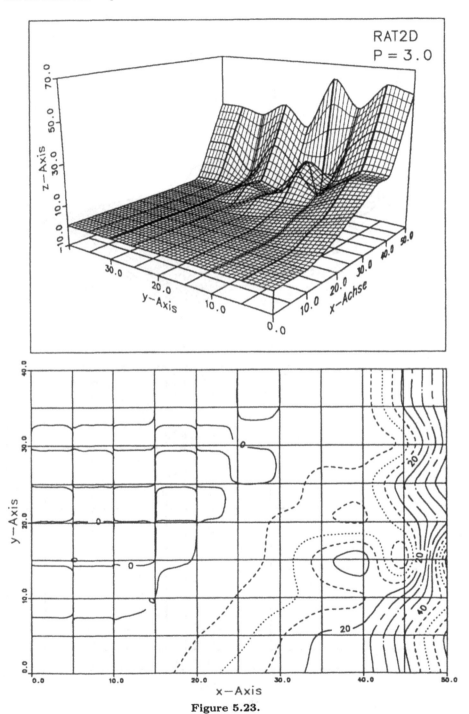

Figure 5.23.

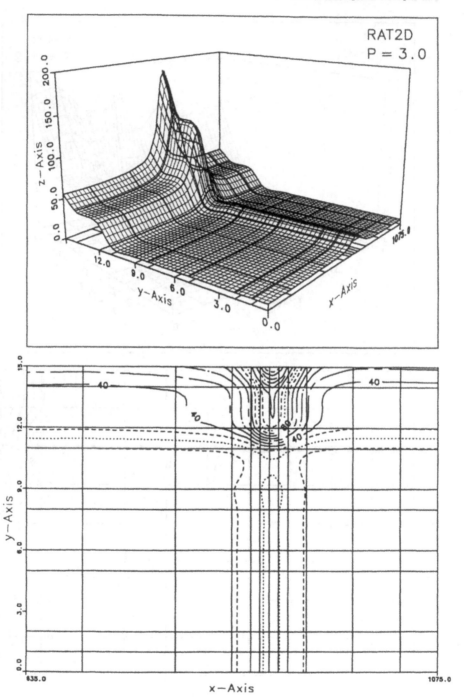

Figure 5.24.

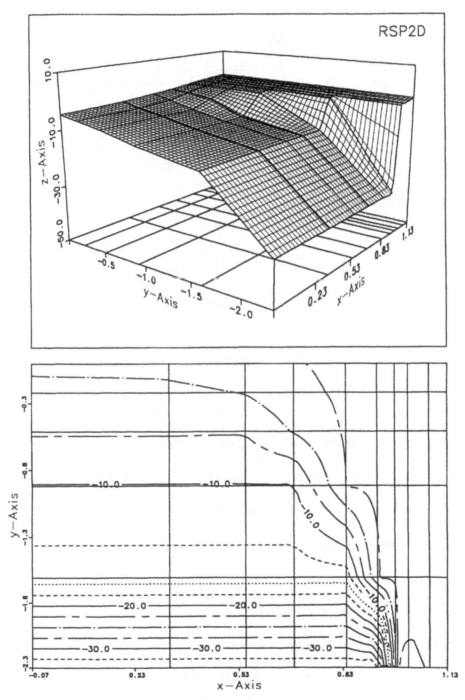

Figure 5.25.

$$h_{i1}(x, x_i, x_{i+1}) \;=\; \frac{1}{\Delta x_i}, \tag{5.14}$$

$$h_{i2}(x, x_i, x_{i+1}, px_i) \;=\; \frac{1}{\Delta x_i}\,\frac{\left(\frac{x_{i+1}-x}{\Delta x_i}\right)^2\left[2px_i\left(\frac{x_{i+1}-x}{\Delta x_i}\right) - 3(1+px_i)\right]}{\left[1 + px_i\left(\frac{x-x_i}{\Delta x_i}\right)\right]^2},$$

$$h_{i3}(x, x_i, x_{i+1}, qx_i) \;=\; -\frac{1}{\Delta x_i}\,\frac{\left(\frac{x-x_i}{\Delta x_i}\right)^2\left[2qx_i\left(\frac{x-x_i}{\Delta x_i}\right) - 3(1+qx_i)\right]}{\left[1 + qx_i\left(\frac{x_{i+1}-x}{\Delta x_i}\right)\right]^2},$$

and consider the form,

$$H(x, y) = (h_{i1}(x), h_{i2}(x), h_{i3}(x)) \begin{pmatrix} b_{11} & b_{12} & b_{13} \\ b_{21} & b_{22} & b_{23} \\ b_{31} & b_{32} & b_{33} \end{pmatrix} \begin{pmatrix} h_{j1}(y) \\ h_{j2}(y) \\ h_{j3}(y) \end{pmatrix}. \tag{5.15}$$

Here, we have suppressed the dependence of the h_{ik} on the additional parameters and we mean

$$\begin{aligned} h_{j1}(y) &= \frac{1}{\Delta y_j}, \\ h_{j2}(y) &= h_{j2}(y, y_j, y_{j+1}, py_j), \\ h_{j3}(y) &= h_{j3}(y, y_j, y_{j+1}, qy_j). \end{aligned}$$

Furthermore, the matrix B depends on the subrectangle R_{ij}, $i = 1, \cdots, n - 1$; $j = 1, \cdots, m - 1$.

Since, by (5.4), $h_{i1} = g'_{i2}$, $h_{i2} = g'_{i3}$, and $h_{i3} = g'_{i4}$, we have

$$H(x, y) = G_{xy}(x, y) \tag{5.16}$$

precisely when B has been chosen to be

$$B = \begin{pmatrix} a_{11} - a_{12} - a_{21} + a_{22} & a_{23} - a_{13} & a_{14} - a_{24} \\ a_{32} - a_{31} & a_{33} & -a_{34} \\ a_{41} - a_{42} & -a_{43} & a_{44} \end{pmatrix}. \tag{5.17}$$

Here, G is the function (4.1), except with the functions g_{i1}, g_{i2}, g_{i3}, and g_{i4} of (5.1), and $A = (a_{ik})$ is the 4×4 coefficient matrix (also depending on i and j) of a C^2 birational spline interpolant.

The first half of the corresponding subroutine RVASPL (Figs. 5.26 and 5.27) is then identical to that of QVASPL. RSP2D is called instead of CUB2D and the coefficients computed from (5.17) instead of from (4.25). (In the program, the h_{ik} are again called g_{ik} and are multiplied by the factor $1/\Delta x_i$. Consequently B, has an additional factor of $\Delta x_i \Delta y_j$.) The FUNCTION RVAVAL (Figs. 5.28 and 5.29) is available for evaluation. It is analogous to QVAVAL and uses the FUNCTION GPIJ (Fig. 5.30).

The examples in Figs. 5.31–5.34 use $px_i = qx_i = py_j = qy_j = p = 3$. They correspond to Figs. 4.29–4.32. The smoothing effect is clearly visible. Since the interpolants tend to the upper surface of the boxes, this method could also be used for rounding off the edges of such boxes. Figures 5.35–5.37 should be compared with Fig. 4.33. In Fig. 5.35, $px_i = qx_i = py_j = qy_j = 3$ again was chosen. In contrast, for Fig. 5.36, we set $px_i = qx_i = 0$ and $py_j = qy_j = 3$ (i.e., quadratics in the x direction instead of rationals). In Fig. 5.37 this was switched to $px_i = qx_i = 3$ and $py_j = qy_j = 0$.

```
     SUBROUTINE RVASPL(N,M,NDIM,MDIM,X,Y,V,U,PX,QX,PY,QY,
    +                  EPS1,EPS2,IR,P,Q,R,A,IFLAG,DX,AX,BX,
    +                  CX,RX,DY,AY,BY,CY,RY)
     DIMENSION X(N),Y(M),V(NDIM,M),U(NDIM,M),PX(N),QX(N),
    +          PY(M),QY(M),P(NDIM,M),Q(NDIM,M),R(NDIM,M),
    +          A(NDIM,MDIM,4,4),DX(N),AX(N),BX(N),CX(N),
    +          RX(N),DY(M),AY(M),BY(M),CY(M),RY(M),
    +          B(4,4),C(4,4),D(4,4),E(4,4)
     IFLAG=0
     IF(N.LT.2.OR.M.LT.2) THEN
          IFLAG=1
          RETURN
     END IF
     N1=N-1
     M1=M-1
     ZERO=0.
     DO 10 I=1,N1
          DX(I)=X(I+1)-X(I)
          U(I,1)=ZERO
10   CONTINUE
     DO 20 J=1,M1
          DY(J)=Y(J+1)-Y(J)
          U(1,J)=ZERO
```

(cont.)

```
20   CONTINUE
     U(N,1)=ZERO
     U(1,M)=ZERO
     DO 50 I=2,N
         DO 40 J=2,M
             H=ZERO
             J1=J-1
             DO 30 K=1,I-1
                 H=H+DX(K)*DY(J1)*V(K,J1)
30           CONTINUE
             U(I,J)=U(I,J1)+H
40       CONTINUE
50   CONTINUE
     IF(IR.EQ.1) GOTO 80
     Q(1,1)=ZERO
     Q(1,M)=ZERO
     DO 60 I=2,N
         I1=I-1
         Q(I,1)=Q(I1,1)+DX(I1)*V(I1,1)
         Q(I,M)=Q(I1,M)+DX(I1)*V(I1,M1)
60   CONTINUE
     P(1,1)=ZERO
     P(N,1)=ZERO
     DO 70 J=2,M
         J1=J-1
         P(1,J)=P(1,J1)+DY(J1)*V(1,J1)
         P(N,J)=P(N,J1)+DY(J1)*V(N1,J1)
70   CONTINUE
80   CALL RSP2D(N,M,NDIM,MDIM,X,Y,U,PX,QX,PY,QY,EPS1,EPS2,IR,P,
    +          Q,R,A,IFLAG,DX,AX,BX,CX,RX,DY,AY,BY,CY,RY)
     DO 120 I=1,N1
         DO 110 J=1,M1
             H=DX(I)*DY(J)
             H1=A(I,J,1,1)-A(I,J,1,2)
             H2=-A(I,J,1,3)
             H3=A(I,J,1,4)
             VZ=1.
             DO 100 K=1,3
                 IF(K.EQ.3) VZ=-VZ
                 K1=K+1
                 A(I,J,K,1)=VZ*H*(A(I,J,K1,2)-A(I,J,K1,1))
                 DO 90 L=2,3
                     A(I,J,K,L)=VZ*H*A(I,J,K1,L+1)
                     VZ=-VZ
90               CONTINUE
100          CONTINUE
             A(I,J,1,1)=A(I,J,1,1)+H1*H
             A(I,J,1,2)=A(I,J,1,2)+H2*H
             A(I,J,1,3)=A(I,J,1,3)+H3*H
110      CONTINUE
120  CONTINUE
     RETURN
     END
```

Figure 5.26. Program listing of RVASPL.

Calling sequence:

CALL RVASPL(N,M,NDIM,MDIM,X,Y,V,U,PX,PY,QX,QY,
              EPS1,EPS2,IR,P,Q,R,A,IFLAG,DX,AX,BX,
              CX,RX,DY,AY,BY,CY,RY)

Purpose:

For given points (x_i, y_j) $i = 1, \cdots, n \geq 2$; $j = 1, \cdots, m \geq 2$ in the plane with $x_1 < \cdots < x_n$ and $y_1 < \cdots < y_m$ and associated heights v_{ij}, this routine determines the coefficients $a_{ijk\ell}$ of a general birational histospline, F, which, on each subrectangle R_{ij}, is of the form (5.15) with h_{ik} (respectively, h_{jk}) according to (5.14), $i = 1, \cdots, n - 1; j = 1, \cdots,$ $m - 1; k = 1, 2, 3$. For IR=1, most of the boundary values needed for this calculation are automatically set to the values of (4.54). However, values for R(1,1), R(N,1), R(1,M), and R(N,M) must be supplied by the user (one could, for example, use the values of (4.56)). For IR=2, all the required boundary values must be supplied by the user.

Description of the parameters:

N,M,NDIM,MDIM,X,Y,PX,QX,PY,QY,EPS1,EPS2,IR,IFLAG,
DX,AX,BX,CX,RX,DY,AY,BY,CY,RY as in RSP2D.

V	ARRAY(NDIM,M): Volume heights v_{ij}, $i = 1, \cdots, n - 1$; $j = 1, \cdots, m - 1$.
A	ARRAY(NDIM,MDIM,4,4): Output: Spline coefficients. A is first used, as dimensioned, as one of the arrays passed to RSP2D. Subsequently, the coefficients of the birational histospline are stored in the subarray A(NDIM,MDIM,3,3).
U,P,Q,R	ARRAY(NDIM,M): Work space.

Required subroutines: RSP2D, RSPDEC, TRDIUA, RSPPER, TRDIUB, RSPMAT.

Figure 5.27. Description of RVASPL.

```
FUNCTION RVAVAL(U,V,N,NDIM,M,MDIM,X,Y,A,PX,QX,PY,QY,IFLAG)
    DIMENSION X(N),Y(M),A(NDIM,MDIM,4,4),PX(N),QX(N),
   +          PY(M),QY(M)
    DATA I,J/2*1/
    IF(N.LT.2.OR.M.LT.2) THEN
        IFLAG=1
        RETURN
    END IF
    CALL INTTWO(X,N,Y,M,U,V,I,J,IFLAG)
    IF(IFLAG.NE.0) RETURN
```

(cont.)
```
   UX=(U-X(I))/(X(I+1)-X(I))
   VY=(V-Y(J))/(Y(J+1)-Y(J))
   PI=PX(I)
   QI=QX(I)
   PJ=PY(J)
   QJ=QY(J)
   RVAVAL=0.
   DO 20 K=1,3
      H=GPIJ(K,UX,PI,QI)
      DO 10 L=1,3
         RVAVAL=RVAVAL+A(I,J,K,L)*H*GPIJ(L,VY,PJ,QJ)
10    CONTINUE
20 CONTINUE
   RETURN
   END
```

Figure 5.28. Program listing of RVAVAL.

FUNCTION RVAVAL(V,W,N,NDIM,M,MDIM,X,Y,A,
                    PX,QX,PY,QY,IFLAG)

Purpose:
Calculation of a function value of a general birational histospline (see the description of RVASPL) at a point $(v, w) \in R$, where R is the underlying rectangular grid.

Description of the parameters:

V,W,N,NDIM,M,MDIM,X,Y,A,PX,QX,PY,QY,IFLAG as in RSPVAL.

Required subroutines: INTTWO, GPIJ.

Figure 5.29. Description of RVAVAL.

```
FUNCTION GPIJ(I,X,PP,QQ)
H=1.-X
H1=PP*X+1.
H2=QQ*H+1.
IF(I.EQ.1) GPIJ=1.
IF(I.EQ.2) GPIJ=H*H*(2.*PP*H-3.*(1.+PP))/(H1*H1)
IF(I.EQ.3) GIPJ=X*X*(2.*QQ*X-3.*(1.+QQ))/(H2*H2)
RETURN
END
```

Figure 5.30. Program listing of GPIJ.

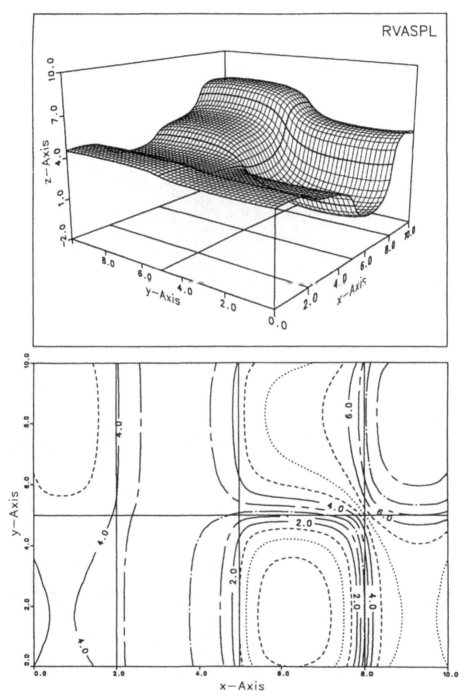

Figure 5.31.

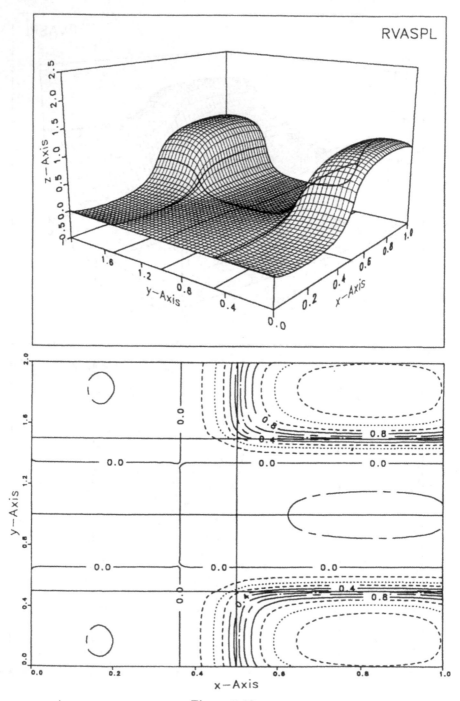

Figure 5.32.

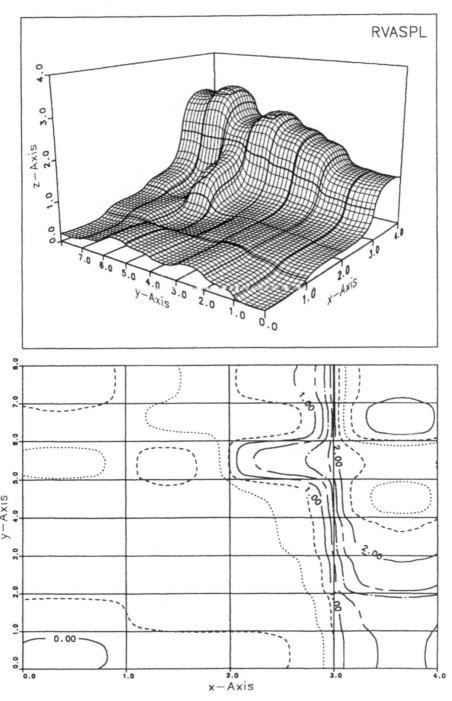

Figure 5.33.

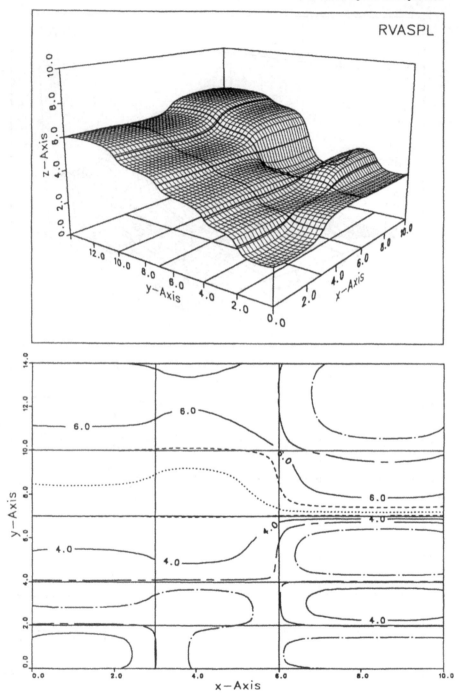

Figure 5.34.

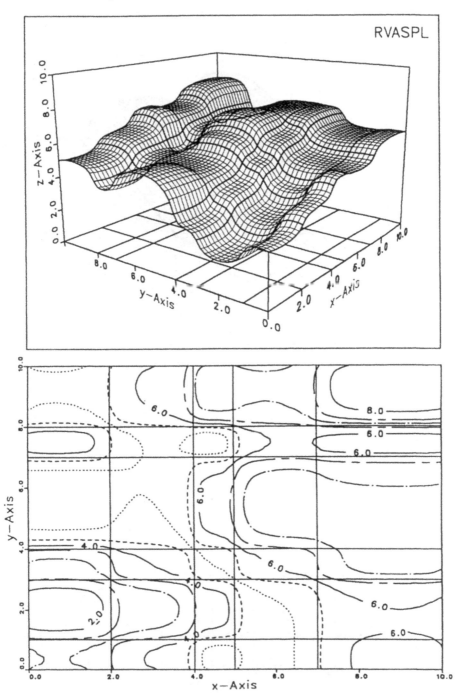

Figure 5.35.

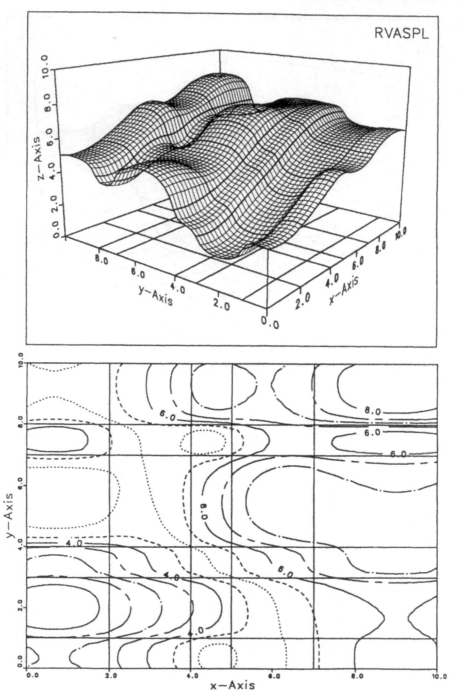

Figure 5.36.

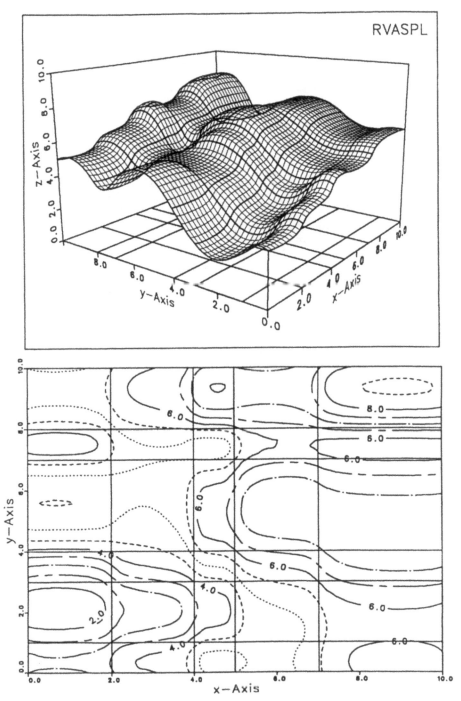

Figure 5.37.

Part II

Spline Interpolation for Arbitrarily Distributed Points

6

Global Methods without Triangulation

6.1. Existence Problems and Goal Setting

In Part II, the interpolation nodes will be allowed to be arbitrarily irregularly distributed points,

$$P_i = (x_i, y_i), \quad i = 1, \cdots, n. \tag{6.1}$$

They need only be pairwise distinct and noncollinear, i.e., not all lying on a single line. Consequently, we must also have $n \geq 3$. This situation arises far more frequently in practical applications than that of Part I. We look for a smooth interpolating function $F = F(x, y)$ with

$$F(x_i, y_i) = z_i, \quad i = 1, \cdots, n, \tag{6.2}$$

where the z_i are the values to be interpolated at the P_i.

As the following example shows, if F were chosen to be a polynomial, then the solvability of the interpolation problem would depend on the degree, the number, and position of the P_i and even the values of the z_i. Suppose that we are given four points $P_1 = (-1, 0)$, $P_2 = (0, -1)$, $P_3 = (1, 0)$, and $P_4 = (0, 1)$, together with values z_1, z_2, z_3, and z_4. Suppose further that we wish to interpolate these values by a bilinear function,

$$F(x, y) = a + bx + cy + dxy.$$

The interpolation conditions (6.2) then give a linear system

$$a - b = z_1,$$
$$a - c = z_2,$$
$$a + b = z_3,$$
$$a + c = z_4,$$

in which d does not appear. There exists a solution for $a, b,$ and c iff the right-hand side is in the image of the corresponding coefficient matrix. This is true precisely when

$$z_1 + z_3 = z_2 + z_4,$$

in which case

$$a = \frac{z_1 + z_3}{2}, \quad b = \frac{z_3 - z_1}{2}, \quad c = \frac{z_1 - 2z_2 + z_3}{2}.$$

The interpolation function obtained in this special case is thus a plane. For arbitrary z_1, z_2, z_3, z_4, this interpolation problem for a bilinear function is not solvable.

In subsequent chapters, we will triangulate the point set, i.e., consider the P_i as vertices of triangles. Triangles that share more than a common vertex must share a common side. In each of these triangles, we will set a function (or several such) of the type,

$$F(x, y) = \sum_{i=0}^{m} \sum_{j=0}^{m-i} a_{ij} x^i y^j = \sum_{i+j=0}^{m} a_{ij} x^i y^j. \qquad (6.3)$$

These functions will then be so arranged that the resulting global function is once continuously differentiable on the set of triangles. For most applications, only one order of smoothness should suffice. These are then *local* C^1 *spline interpolation methods*. For the usual reasons, we consider only low-degree polynomials, i.e., linear ($m = 1$, only C^0), quadratic ($m = 2$), and cubic ($m = 3$) polynomials. We will also very briefly discuss the cases, $m = 4$ and $m = 5$. We do not use bilinear, biquadratic, or bicubic polynomials. The number of coefficients here is only $(m + 1)(m + 2)/2$, i.e., 3, 6, 10, 15 and 21 for $m = 1, \cdots, 5$. Since this number of coefficients is, for $m > 1$, too high in comparison to the three interpolation conditions per triangle, it will be necessary to estimate values for partial derivatives at the P_i. The mathematical background needed for local C^1 spline interpolation involves many complicated formulas and so we will often not give many details in this regard.

All possible local and global methods are described and compared by examples in the survey articles [7,28,37,41,93] as well as in the book [58]. In this chapter, we will only give two very simple global methods. These can be very easily programmed and often give results just as visually pleasing as local C^1 spline interpolation does. However, for numerical reasons, they cannot be used on larger point sets ($n \approx 100$). In contrast, the number of points plays no role in a local method.

6.2. Shepard's Method

The simplest method is Shepard's method, which is based on the following considerations ([93]). For given weights $w_i > 0$, we look for the value S that best approximates the z_i, $i = 1, \cdots, n$, in the sense of weighted least squares, i.e., we minimize

$$\sum_{i=1}^{n} w_i(z_i - S)^2$$

over S. This minimizer is easily computed to be

$$S = \frac{\sum_{i=1}^{n} w_i z_i}{\sum_{i=1}^{n} w_i}.$$

Shepard's method ([97]) is now obtained by choosing the weights w_i to vary as an inverse power of the Euclidean distance from an arbitrary point $P = (x, y)$ to $P_i = (x_i, y_i)$, i.e., by setting

$$w_i = w_i(x, y) = \frac{1}{r_i^q} \quad (q > 0), \tag{6.4}$$

where

$$r_i = \sqrt{(x - x_i)^2 + (y - y_i)^2}. \tag{6.5}$$

The greater the distance from P to P_i, the smaller the influence of P_i on the values of

$$S_0(x, y) = \sum_{i=1}^{n} z_i v_i(x, y), \tag{6.6}$$

where

$$v_i(x, y) = \frac{w_i(x, y)}{\sum_{k=1}^{n} w_k(x, y)} = \frac{r_i^{-q}}{\sum_{k=1}^{n} r_k^{-q}}. \tag{6.7}$$

In place of (6.7), we will use the numerically more stable form,

$$v_i(x, y) = \frac{\prod_{j \neq i} r_j^q}{\sum_{k=1}^n \prod_{j \neq k} r_j^q} \tag{6.8}$$

in (6.6).

Since

$$v_i(x_j, y_j) = \delta_{ij}, \tag{6.9}$$

S_0 interpolates, i.e.,

$$S_0(x_i, y_i) = z_i, \quad i = 1, \cdots, n. $$

Further, from

$$v_i(x, y) \geq 0, \quad \sum_{k=1}^n v_k(x, y) = 1, \tag{6.10}$$

follows the useful property that

$$\min_i z_i \leq S_0(x, y) \leq \max_i z_i. \tag{6.11}$$

In particular,

$$z_i \geq 0, \quad i = 1, \cdots, n \quad \Rightarrow \quad S_0(x, y) \geq 0, \tag{6.12}$$

and

$$z_i = c, \quad i = 1, \cdots, n \quad \Rightarrow \quad S_0(x, y) = c. \tag{6.13}$$

The functions $v_i = v_i(x, y)$ are everywhere infinitely often differentiable except at the points P_i, where all partial derivatives up to order $q - 1$ if q is an integer and up to $[q]$ if q is not an integer vanish. For $q = 1$, the one-sided partials of v_i do exist at P_i; however, their values are almost always different. If $q < 1$, then there do not exist any partial derivatives ([47]). Hence, S_0 is likewise infinitely often differentiable except at the points P_i, where it could only be continuous.

In particular, for the often suggested choice of $q = 2$, we have

$$q = 2 \quad \Rightarrow \quad \left. \frac{\partial S_0}{\partial x} \right|_{P_i} = \left. \frac{\partial S_0}{\partial y} \right|_{P_i} = 0, \tag{6.14}$$

from which it follows that the surface has the appearance of being flat in a neighborhood of the P_i, $i = 1, \cdots, n$. This effect is evident, for example, in Fig. 6.3. We will further comment on this particular figure later. The root cause of this phenomenom is that S_0 involves only the length of the difference vector between P and P_i and not its direction ([97]). This effect can be eliminated when approximate values for the first partial derivatives,

$$zx_i = \left.\frac{\partial S_0}{\partial x}\right|_{P_i}, \quad zy_i = \left.\frac{\partial S_0}{\partial y}\right|_{P_i}, \quad i = 1, \cdots, n \quad (6.15)$$

are available from one source or another.

The function,

$$S_1(x, y) = \sum_{i=1}^{n} [z_i + (x - x_i)zx_i + (y - y_i)zy_i] v_i(x, y), \quad (6.16)$$

besides having the same properties as S_0 ([29]), also takes on the given values (6.15) for the first partials.

S_1 with variable $q > 0$ is implemented in the subroutine SHEPG (Figs. 6.1 and 6.2). In order to avoid overflows, each r_i is divided by RMAX $=$ $\max_i r_i$. This may result in underflows, but these are detected and signaled by IFLAG=2. If before calling SHEPG, all the entries of the array ZXZY, which is to contain the estimates of the partial derivatives, are set to zero, then the function S_0 of (6.6) is computed instead of S_1.

Although we could, of course, apply SHEPG to the examples of Part I, we choose an example with irregularly distributed data. The exact location of the points can be seen in the lower half of Fig. 6.3; the z_i are marked by heavy dots in the upper half. Figure 6.3 shows the surface resulting from S_0 (SHEP) and $q = 2$. The unpleasant property (6.14) is clearly visible. Figures 6.4 and 6.5 show the results for S_1 with $q = 2$ and $q = 4$. The required values of the first partials were estimated using the subroutine GRADL. We will return to this program in Chapter 8. The improvement of Fig. 6.4 over 6.3 is rather evident. Figure 6.5 shows that q should not be chosen to be too large. On the basis of other examples that we computed, we suggest (also for computational reasons as r_i^2 requires no square root) to always take $q = 2$ and in any case to at least avoid $q < 1.5$ and $q > 3$. Otherwise, (but also already for $q = 2$), numerical instabilities arise for large n that make a correct evaluation of S_1 numerically impossible.

Numerous variants of Shepard's method have been discussed in the literature. It is also possible to specify the values of higher-order partial derivatives ([7,29]). Other metrics can be used instead of Euclidean distance; these even may be different for each point. For example, with a fixed metric, it is possible to consider a different exponent for each point ([47]). There are local variants ([93,97]) for which only the points P_i in a neighborhood of the point of evaluation, P, have an influence on the function value. However, these guarantee only once continuous differentiability (except at the P_i). It is also possible to consider better functions v_i such as, for example ([7,41]),

$$v_i(x, y) = \left[\frac{(R - r_i)_+}{Rr_i} \right]^2, \qquad (6.17)$$

with $R > 0$. Finally, the z_i in (6.6) may be replaced by a function Q_i with $Q_i(x_i, y_i) = z_i$; this and a variable parameter R are suggested, for example, in [86]. The subroutine QSHEP2D([87]) is a professional implementation of this idea. There is even a three-dimensional version, QSHEP3D([88]).

```
     FUNCTION SHEPG(N,U,V,Q,X,Y,Z,ZXZY,R,IEMIN,IFLAG)
     DIMESNION X(N),Y(N),Z(N),ZXZY(2,N),R(N)
     IFLAG=0
     IF (Q.LE.0.0) THEN
         IFLAG=1
         RETURN
     END IF
     EMIN=REAL(IEMIN)
     RMAX=0.0
     DO 10 J=1,N
         DU=U-X(J)
         DV=V-Y(J)
         RJ=DU*DU+DV*DV
         IF (Q.NE.2.0) RJ=SQRT(RJ)**Q
         R(J)=RJ
         IF (RJ.GT.RMAX) RMAX=RJ
10   CONTINUE
     SUM1=0.0
     SUM2=0.0
     DO 30 I=1,N
         PROD=1.0
         DO 20 J=1,N
             IF (J.NE.I) THEN
                 RJ=R(J)/RMAX
                 IF (RJ.EQ.0.0) GOTO 30
                 EXP=-LOG10(PROD)-LOG10(RJ)
                 IF (EXP.GT.EMIN) THEN
                     IFLAG=2
                     RETURN
                 END IF
                 PROD=PROD*RJ
             END IF
20       CONTINUE
         SUM1=SUM1+(Z(I)+(U-X(I))*ZXZY(1,I)
    &                   +(V-Y(I))*ZXZY(2,I))*PROD
         SUM2=SUM2+PROD
30   CONTINUE
     SHEPG=SUM1/SUM2
     RETURN
     END
```

Figure 6.1. Program listing of SHEPG.

FUNCTION SHEPG(N,U,V,X,Y,Z,ZXZY,R,IEMIN,IFLAG)

Purpose:
For given points (x_i, y_i), $i = 1, \cdots, n$, arbitrarily distributed in the plane, and associated heights z_i, Shepard's interpolant, including the first partial derviatives with respect to x and y at the nodes, is evaluated at the point (u, v).

Description of the parameters:

N	Number of given points.
U	x coordinate of the point of evaluation.
V	y coordinate of the point of evaluation.
Q	Value of the exponent q to which the distances r_i from (u, v) to (x_i, y_i), $i = 1, \cdots, n$, are raised in Shepard's method. Restriction: Q > 0.
X	ARRAY(N): x values.
Y	ARRAY(N): y values.
Z	ARRAY(N): z values.
ZXZY	ARRAY(2,N): Partial derivatives with respect to x and y at the nodes. The first component must contain the partial derivatives with respect to x and the second component those with respect to y. Values for these can, for example, be computed using the subroutines GRADG, GRADL, GRADH, GRADLK, or GRADLA.
R	ARRAY(N): Work space. The values of the r_i^q are stored here.
IEMIN	Input: largest negative exponent for a real constant in floating-point representation.
IFLAG	=0: Normal execution.
	=1: Q ≤ 0.
	=2: Underflow detected.

Figure 6.2. Description of SHEPG.

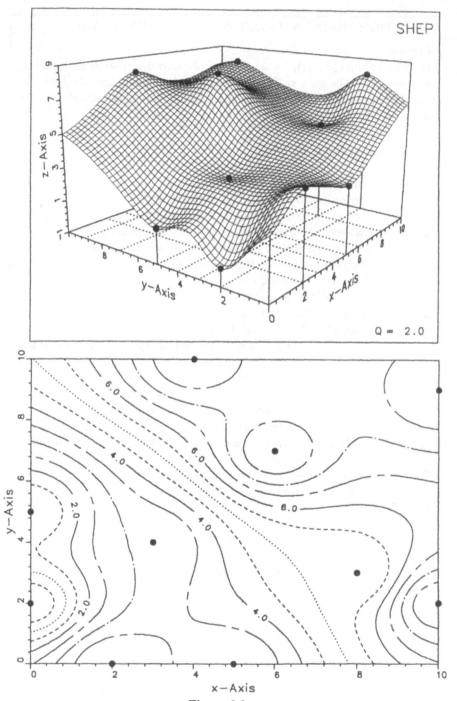

Figure 6.3.

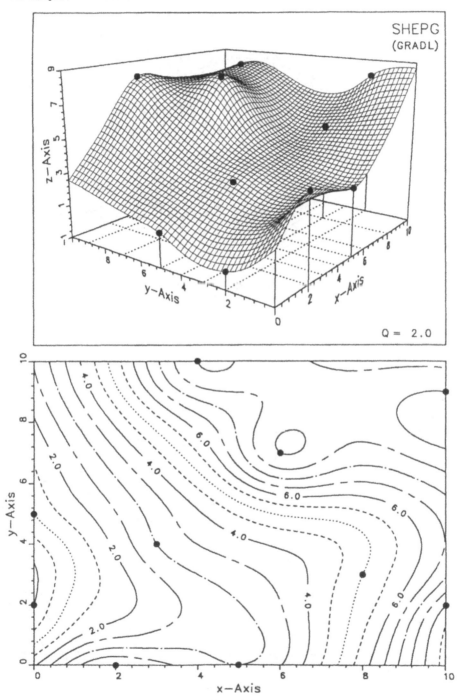

Figure 6.4.

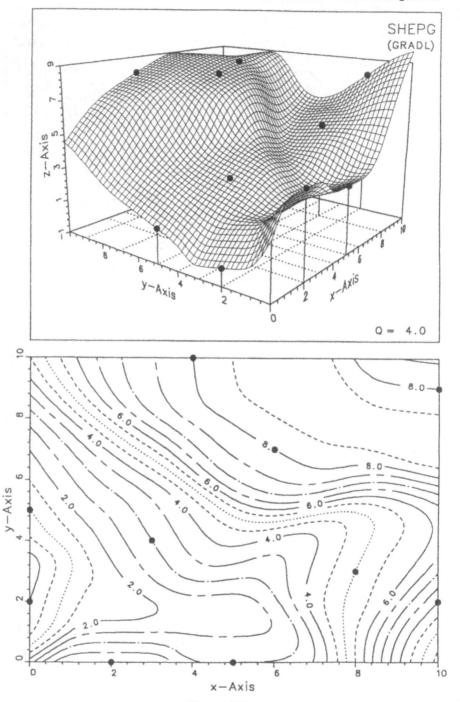

Figure 6.5.

6.3.   Hardy's Multiquadrics

We now consider an interpolating function of the form,

$$H(x,y) = \sum_{k=1}^{n} a_k B_k(x,y). \tag{6.18}$$

Here, the B_k are certain given basis functions and the a_k are to be determined so that the interpolation conditions,

$$\sum_{k=1}^{n} a_k B_k(x_i, y_i) = z_i, \quad i = 1, \cdots, n \tag{6.19}$$

are sastisfied. Of course, the basis functions of this global interpolation method are to be chosen so that H looks as smooth as possible and that the linear system (6.19) is as well-conditioned as possible ([34]), i.e., in particular, also so that it has a solution (which can be calculated in a numerically stable way).

In this regard, the so-called radial basis functions,

$$B_k(x,y) = B_k(r_k), \quad r_k = \sqrt{(x-x_k)^2 + (y-y_k)^2}, \tag{6.20}$$

have proven themselves to be quite useful. Some possibilities include ([51,$q = 1/2$])

$$B(r_k) = (r_k^2 + R^2)^q, \quad q \in \mathbb{R}, \tag{6.21}$$

and also ([26,28,41])

$$\begin{aligned} B(r_k) &= \ln(r_k^2 + R^2), \\ B(r_k) &= (r_k^2 + R^2)^m \ln(r_k^2 + R^2), \quad m > 0. \end{aligned}$$

Here, R is a parameter that should not be chosen to be too large. In any case, the coefficient matrix of (6.19) is full and, since

$$B_k(x_i, y_i) = B_i(x_k, y_k), \tag{6.22}$$

symmetric. Strict diagonal dominance or positive definiteness are not realizable. As already mentioned, for practical reasons the condition number of the matrix should not be too large ($> 1.\text{E6}$).

Here, we limit ourselves to the simplest form (6.21). Once R and q have been chosen, we form the matrix with entries (6.22). Finally, the linear system (6.19) is solved numerically, in this case by DECOMP and SOLVE ([34]). These do not make use of the symmetry (6.22), but they do give an estimate COND of the condition number of the matrix ([34]). If no

$R > 0$ is given in the corresponding subroutine HARDY (Figs. 6.6 and 6.7), following [102], R is automatically set to

$$R = \sqrt{\frac{1}{10} \max \left(\max_{i,k} |x_i - x_k|, \max_{i,k} |y_i - y_k| \right)} \qquad (6.23)$$

and then rounded to one decimal place. This R is used for all the following examples; as COND increases with R, substantially larger values should be avoided. The FUNCTION FHARDY (Figs. 6.8 and 6.9) may be used to evaluate (6.18).

The examples in Figs. 6.10–6.12 (cf. Fig. 4.9) and 6.13–6.15 (cf. Fig. 4.10) have the same initial data but differ in the choice of $q = -1, -1/2, 1/2$. The

```
      SUBROUTINE HARDY(NDIM,N,Q,R,X,Y,Z,A,B,IPVT,COND,WORK)
      DIMENSION X(N),Y(N),Z(N),A(N),B(NDIM,N),IPVT(N),WORK(N)
      IF (R.LE.0.0) THEN
         X1=X(1)
         XN=X(1)
         Y1=Y(1)
         YN=Y(1)
         DO 10 I=2,N
            XI=X(I)
            YI=Y(I)
            IF (XI.LT.X1) X1=XI
            IF (XI.GT.XN) XN=XI
            IF (YI.LT.Y1) Y1=YI
            IF (YI.GT.YN) YN=YI
10       CONTINUE
         X1=XN-X1
         Y1=YN-Y1
         R=INT(10.0*SQRT(AMAX1(X1,Y1)/10.))/10.0
      END IF
      R2=R*R
      DO 30 I=1,N
         A(I)=Z(I)
         B(I,I)=R2**Q
         XI=X(I)
         YI=Y(I)
         DO 20 J=I+1,N
            B(I,J)=((X(J)-XI)**2+(Y(J)-YI)**2+R2)**Q
            B(J,I)=B(I,J)
20       CONTINUE
30    CONTINUE
      CALL DECOMP(NDIM,N,B,COND,IPVT,WORK)
      CALL SOLVE(NDIM,N,B,A,IPVT)
      RETURN
      END
```

Figure 6.6. Program listing of HARDY.

Calling sequence:

CALL HARDY(NDIM,N,Q,R,X,Y,Z,A,B,IPVT,COND,WORK)

Purpose:
For points (x_i, y_i), $i = 1, \cdots, n$, arbitrarily distributed in the plane, and associated heights z_i, the coefficients a_i of the Hardy multiquadric interpolant are calculated.

Description of the parameters:

NDIM	Maximum first dimension of B. Restriction: NDIM $\geq$ N.
N	Number of given points.
Q	Value of the exponent q in Hardy's method.
R	Value of the quantity R in Hardy's method. If an $R \leq 0.0$ is given, then R is set to SQRT(AMAX1(XD,YD)/10.) (with one decimal point). Here, XD is the difference between the largest and smallest x values. YD is analogous for the y values.
X	ARRAY(N): x-coordinates of the nodes.
Y	ARRAY(N): y-coordinates of the nodes.
Z	ARRAY(N): Heights z_i.
A	ARRAY(N): Output: Coefficients a_i of the interpolant.
COND	Output: Estimate of the condition number of the associated linear system.
B	ARRAY(NDIM,N): Work space.
IPVT, WORK	ARRAY(N): Work space.

Required subroutines: DECOMP, SOLVE.

Figure 6.7. Description of HARDY.

```
      FUNCTION FHARDY(N,Q,R,X,Y,A,U,V)
      DIMENSION X(N),Y(N),A(N)
      FHARDY=0.0
      R2=R*R
      DO 10 I=1,N
          FHARDY=FHARDY+A(I)*((U-X(I))**2+(V-Y(I))**2+R2)**Q
10    CONTINUE
      RETURN
      END
```

Figure 6.8. Program listing of FHARDY.

FUNCTION FHARDY(N,Q,R,X,Y,A,U,V)

Purpose:
For points (x_i, y_i), $i = 1, \cdots, n$, arbitrarily distributed in the plane, and associated heights z_i, the Hardy multiquadric interpolant is evaluated at the point (u, v).

Description of the parameters:

N,Q,R,X,Y as in HARDY.
A   ARRAY(N): The coefficients of the interpolant as calculated
           by HARDY.
U x-coordinate of the point of evaluation.
V y-coordinate of the point of evaluation.

Figure 6.9. Description of FHARDY.

$q = -1$ and $q = -1/2$ plots are out of the question; the exponent $q = 1/2$ is almost always preferable. Such a statement depends also on the the value of R. If R were set to $R = .2$ then Figs. 6.13 and 6.14 would appear very different from 6.15. On the other hand, the estimate COND of the condition number increases rapidly with q. For the example in Fig. 6.10, we obtained the estimates 1.59, 5.21, 9.98, and 5.74E7 for $q = -1, -1/2, 1/2, 1$. For $q = 1$, almost all the examples we considered had COND $> 1.\text{E}8$ and consequently its use should not be considered. Perhaps it is advisable to choose q somewhat smaller than $1/2$.

For larger values of n, the method sometimes does not work numerically, presumably because of a large condition number. For the data of Fig. 2.9 (cf. Figs. 4.13, 5.20) with $n = 121$ and for that of Fig. 2.12 (cf. Figs. 3.10, 4.16, 5.23) with $n = 99$, we obtained for $q = -1/2, 1/2$, very smooth-looking surfaces without having any numerical problems. The largest condition number encountered was with COND$=3.7\text{E}8$ for $n = 121$ and $q = 1/2$. The results for the data of Fig. 2.11 (cf. Figs. 4.15, 4.24, 5.22, 5.25) with $n = 60$ and for that of Fig. 2.13 (cf. Figs. 3.11, 3.25, 4.17, 4.25, 5.24) were rather unusable. For $q = -1/2$, the condition numbers were 2.48E8 and 1.08E6, and for $q = 1/2$, they were 6.32E9 and 1.35E9.

For larger and larger values of n, better and better numerical methods must be used ([25,26]). Nevertheless, it is doubtful that for such n these problems can be successfully dealt with. Large values of n present no such problems for local C^1 spline methods (over rectangular grids or irregularly distributed points).

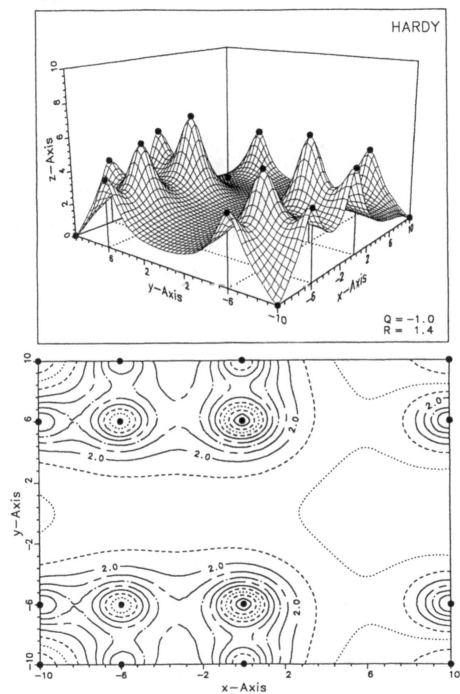

Figure 6.10.

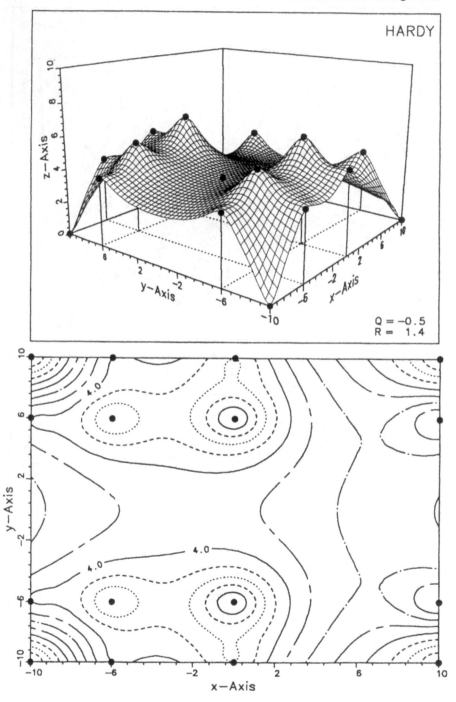

Figure 6.11.

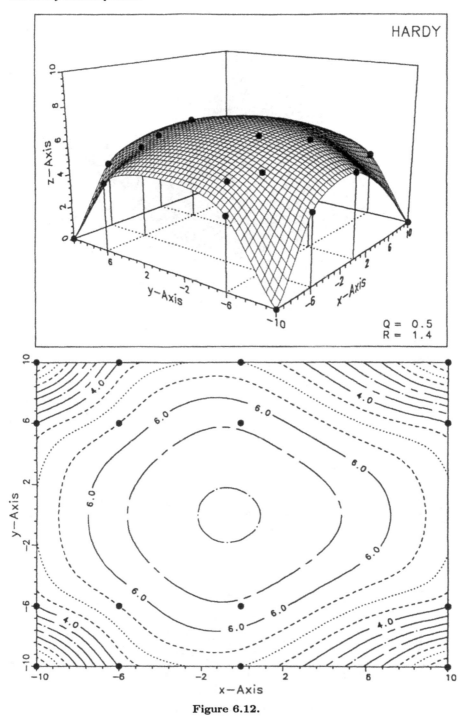

Figure 6.12.

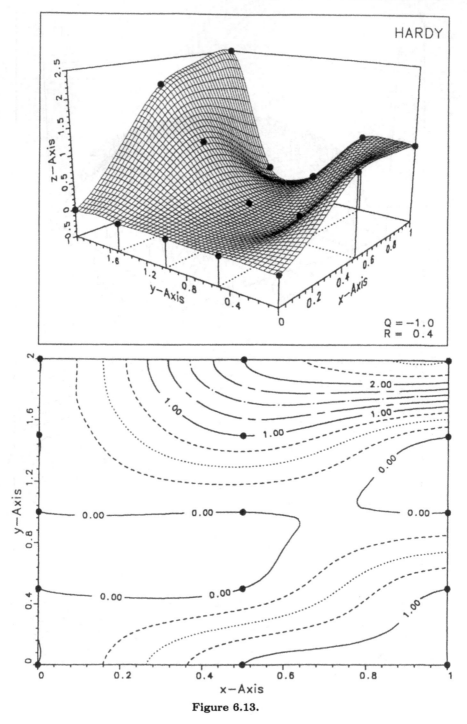

Figure 6.13.

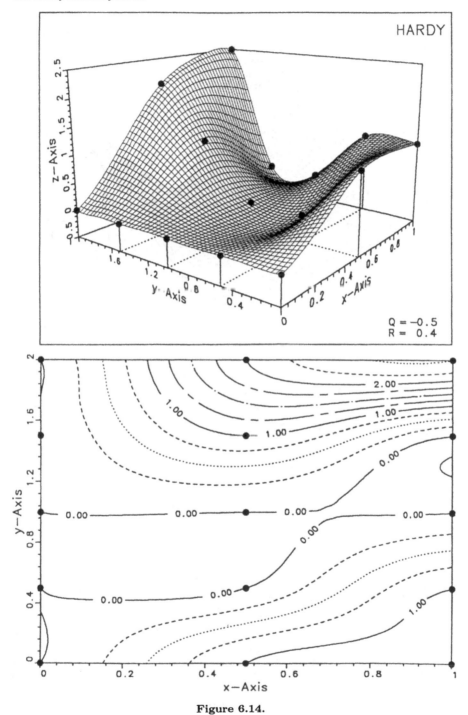

Figure 6.14.

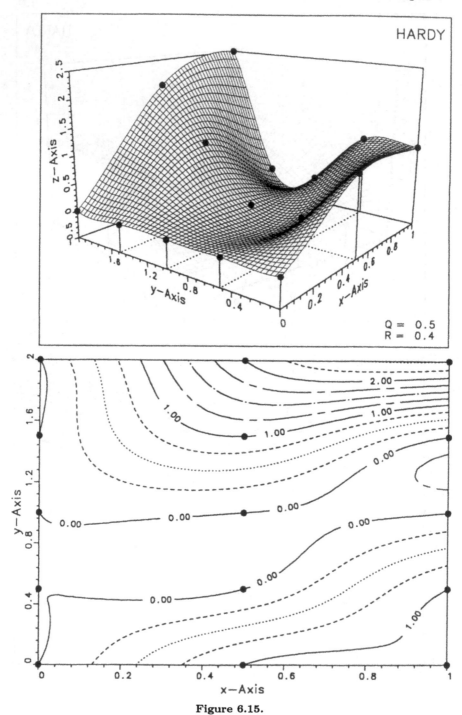

Figure 6.15.

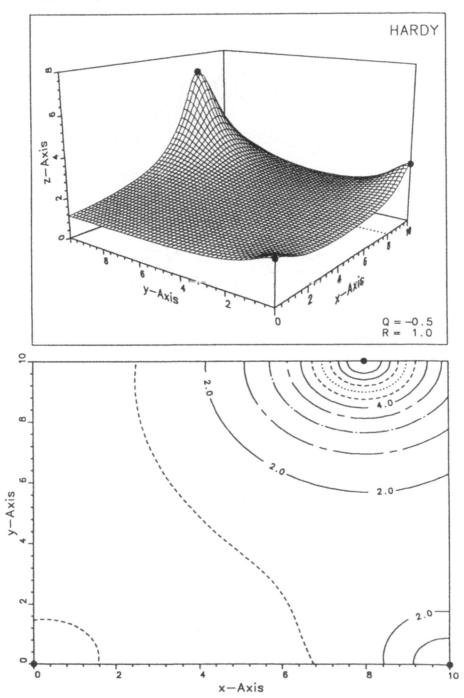

Figure 6.16.

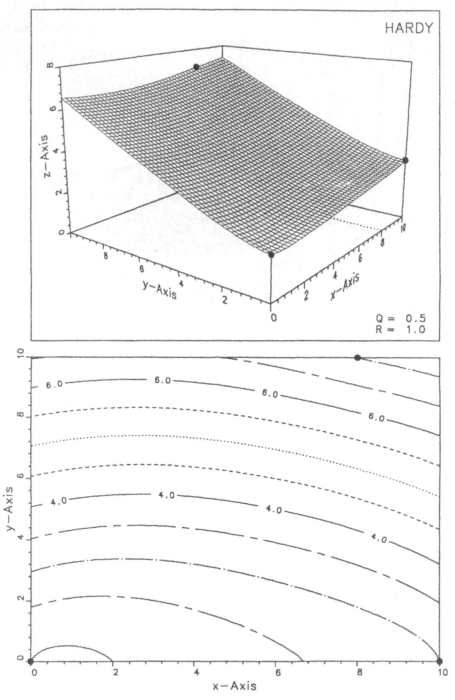

Figure 6.17.

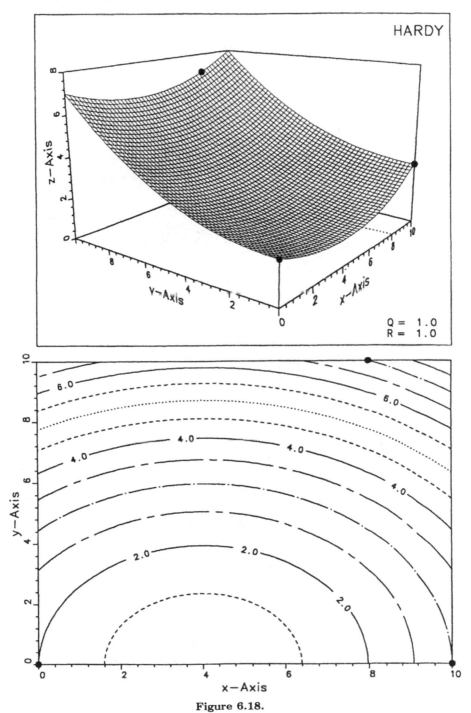

Figure 6.18.

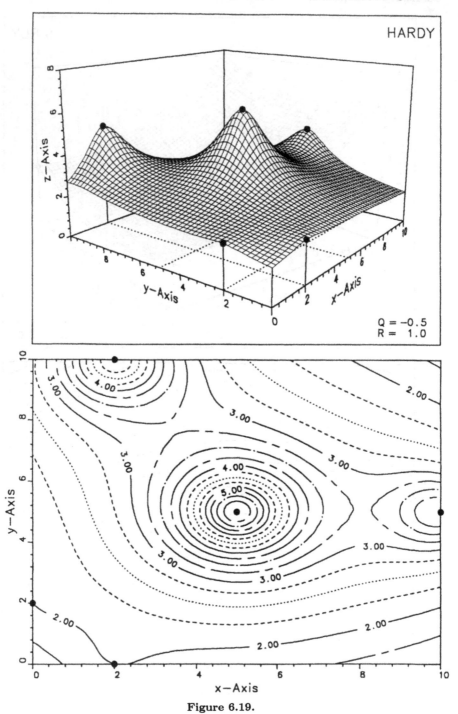

Figure 6.19.

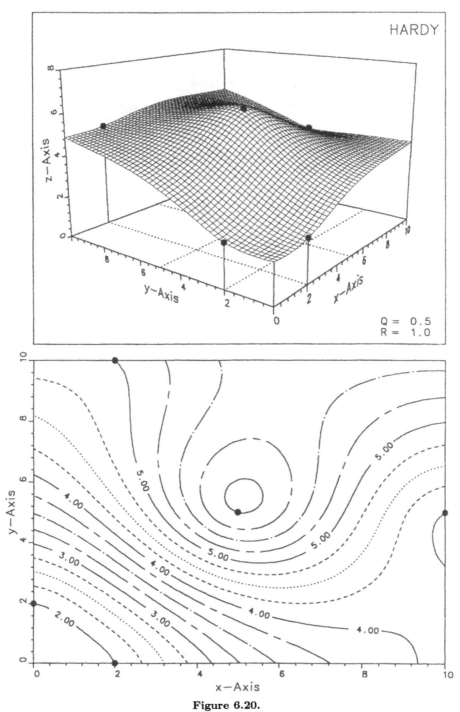

Figure 6.20.

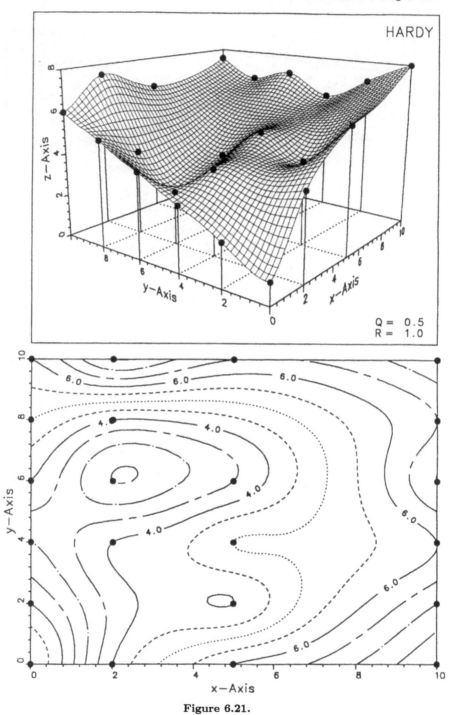

Figure 6.21.

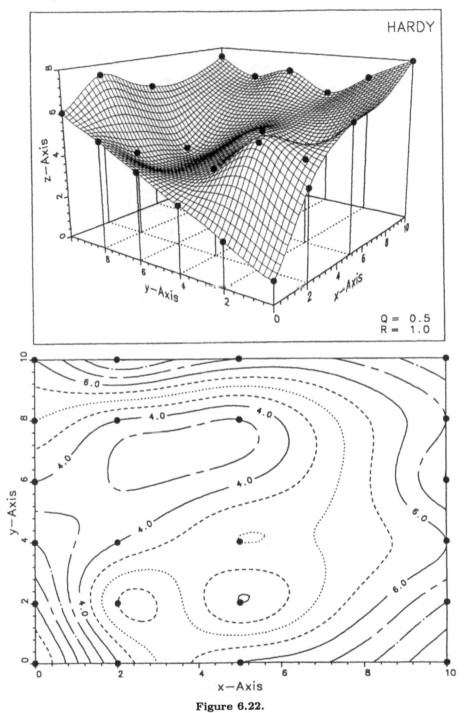

Figure 6.22.

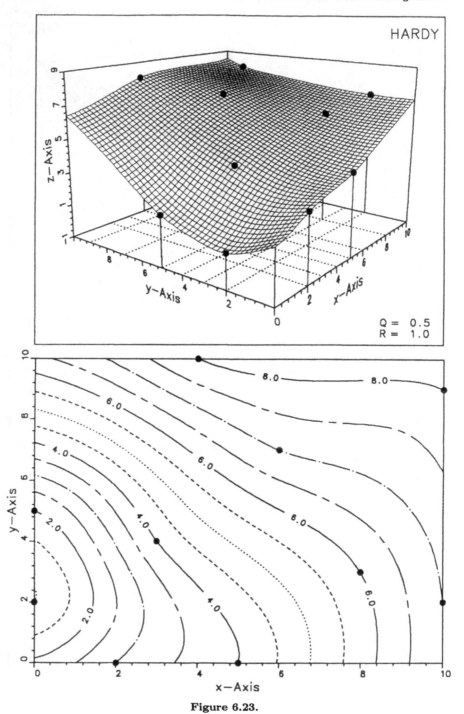

Figure 6.23.

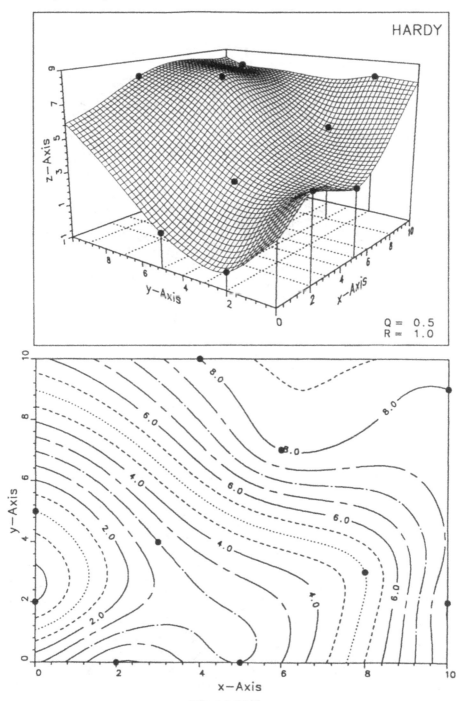

Figure 6.24.

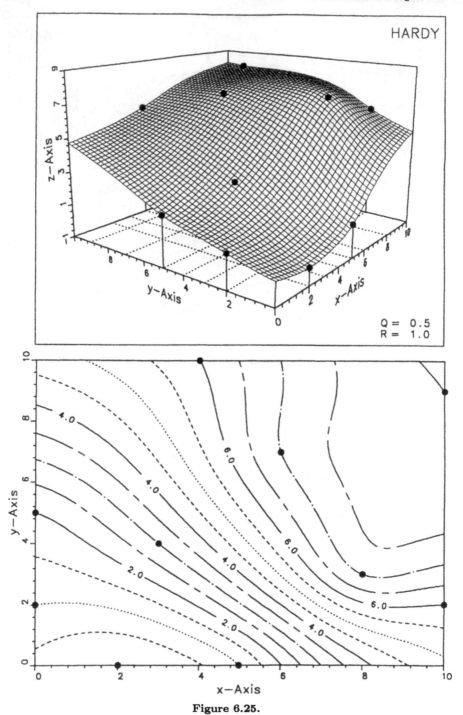

Figure 6.25.

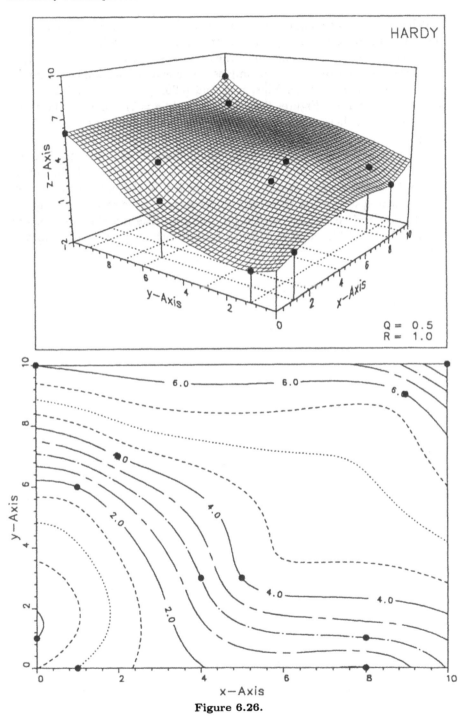

Figure 6.26.

While in the preceding examples the points P_i lie on a rectangular grid, the following ones use irregularly distributed P_i. We will also return to these later. Figures 6.16–6.18 show the same example for $q = -1/2, 1/2, 1$. This was the only one of the examples we considered for which we could get a result for $q = 1$. In Figs. 6.19 and 6.20, $q = -1/2$ and $q = 1/2$ are again compared. Figures 6.21–6.26 use only $q = 1/2$. Due to the small pointsets, in all these cases (Figs. 6.16–6.26), COND did not exceed 2.84E2 for $q = -1$, $q = -1/2$, or $q = 1/2$.

Altogether, Hardy's method is very promising ([41,56]). How to choose a good R depending on n and the z_i or whether to introduce point-dependent values R_k are still open questions. The problem of the numerical solution of the linear systems that arise has not, for large n, been solved to complete satisfaction and could also remain problematic. The modified methods of Maude([35]) and McLain([35,36]) are similar, but not particularly successful.

7

Triangulations

As we already mentioned at the beginning of the previous chapter, local C^1 spline interpolation requires a suitable *triangulation* of the set $P = \{P_1, \cdots, P_n\}$ of interpolation nodes. We will assume that these are pairwise distinct and do not all lie on a single line. The explanations that directly follow are based on [95], where there are also further references to the literature besides those that we cite here.

A triangulation T is a set of N triples of points (P_i, P_j, P_k) with $i, j, k \in \{1, \cdots, n\}$ (i, j, k pairwise distinct) such that for each triple, the corresponding points are the vertices of a triangle with the properties that each such triangle contains only those three points of P and those as vertices, that the intersection of the interiors of any two such triangles is empty, and that the union of all the triangles is the convex hull of P. A mathematically more elegant definition, although not as clear, is that a triangulation is a maximal subset of the set of all edges with endpoints in P that have only vertices in common ([14]).

For any set of interpolation nodes, there are almost always several triangulations. Figure 7.1, which we shall also later use for other purposes, shows two different triangulations. In this special case, however, they differ only in the subdivision of a certain convex quadrilateral. For N, the number of triangles and, K, the number of sides (edges), it is always true ([14,60]) that

$$N = 2n - n_R - 2 \tag{7.1}$$

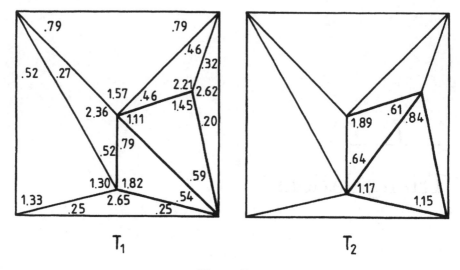

$$T_1 \qquad\qquad\qquad\qquad T_2$$

Figure 7.1.

and

$$K = 3n - n_R - 3, \tag{7.2}$$

where n_R is the number of points on the boundary of the convex hull of
P. Property (7.1) implies that the number of triangles in a triangulation
depends only on P, and hence we may without ambiguity speak of the
number of point triples of a triangulation. Further, N and K also satisfy
([17])

$$n - 2 \le N \le 2n - 5 \tag{7.3}$$

and

$$2n - 3 \le K \le 3n - 6. \tag{7.4}$$

As there is almost always more than one triangulation of any set P, we
need a criterium in order to select that triangulation most suitable for the
purposes of interpolation as well as an algorithm by which to find it. For
purposes of motivation, instead of a general set P, we consider one of four
points that are the vertices of a convex quadrilateral. In this case, there are
always exactly two possible triangulations: labeling the points as P_1, P_2,
P_3, and P_4 e.g., counterclockwise, then we may choose either the diagonal
$\overline{P_1P_3}$ or the diagonal $\overline{P_2P_4}$.

One could perhaps decide that the best triangulation is the one for which
the diagonal has the shortest length. Unfortunately, this often results in
long, skinny, triangles which are to be avoided ([94,95]) in interpolations.
Hence, what we do is find, for both triangulations, the smallest angle in each
of the two triangles, and say that the triangulation for which the smallest

of the two smallest angles is largest is the better one. This is the case for
the triangulation T_2 of the heavy-lined quadrilateral shown in Fig. 7.1; the
triangulation T_1 of the same quadrilateral is poorer. The definition handles
the possibility of both triangles having the same size angles. In general, the
triangulation with diagonal $\overline{P_1P_3}$ is in the preceding sense better than that
with $\overline{P_2P_4}$ precisely when P_4 is inside the circle through P_1, P_2, and P_3.
In special cases, the two triangulations may be equally good. Naturally,
there are also other criteria ([8,70,94,95]). We will not go into these, as for
reasons that will soon be apparent, we have decided in this book to use only
the previous criterium. We now proceed to generalize this to larger node
sets.

We will say that a *triangulation* T of $P = \{P_1, \cdots, P_n\}$, $n > 4$, is *locally
optimal* if each of its convex quadrilaterals has been optimally triangulated
according to the previous criterium. If we wish to compare two different
triangulations, which is to be best must be decided by making an evaluation
of all the triangles defined by the respective triangulations. Such a global
criterium is obtained in the following manner ([60]) To each triangulation
T, we assign a vector $a = a(T) = (a_1, \cdots, a_N)$ whose components are the
values of the smallest angles (in radians) of all N triangles, arranged in
increasing order. Two such vectors, a and b (and hence their associated
triangulations), can be lexicographiclly ordered: $a < b$ means $a_i = b_i$,
$i = 1, \cdots, m - 1$, but $a_m < b_m$ for some $m \leq N$. A *triangulation* is then said
to be *optimal* if its associated vector is lexicographically at least as large
as those for all other triangulations. Such a triangulation is also called a
Thiessen or *Delaunay triangulation*. For example, in Fig. 7.1 ([70]),

$$a(T_1) = (.20, .25, .27, .46, .52, .54, .59, .79)$$

and

$$a(T_2) = (.20, .25, .27, .46, .52, .61, .79, .84).$$

Hence, T_2 is better than T_1 (T_2 is even an optimal triangulation). Such
optimal triangulations always exist, since the number of all possible triangulations is, of course, finite. Every *optimal* triangulation is also *locally
optimal*, since if such a triangulation were not locally optimal, one could
switch the two diagonals in a convex quadrilateral and thereby lexicographically increase the associated vector. A very essential property of the previously described evaluation criterium that, in contrast, does not hold for
other known criteria is that each *locally optimal* triangulation is also *optimal* ([95]). This is ultimately the basis for the construction of algorithms
for finding *optimal Delaunay triangulations*.

In principle, there are three different algorithms. In Algorithm A, one
starts from almost any triangulation and then in all possible quadrilaterals
tries exchanging diagonals to see if the associated vector increases. In

algorithm B, one starts with any triangle that has no nodes in its interior and then adds points one at a time so that the resulting triangulations are always locally optimal. Lastly, in Algorithm C, P is split into two parts, each part is triangulated locally optimally, and then the triangulations merged so that they stay locally optimal. Algorithms A and B require at least $O(n^2)$ operations. Under a certain assumption, algorithm C can be carried out in $O(n \log(n))$ operations.

All the algorithms can be implemented with differing storage requirements, which also have an effect on execution speed. Since a triangulation is completely described by the listing of N index triples, this way of recording the triangulation would require at least $3N$ storage locations. One could also choose to describe the triangulation by listing the $2K$ index pairs corresponding to the two vertices of each edge of each triangle. The dimensioning of the arrays requires the upper bounds (7.3) and (7.4). Efficient memory utilization techniques are described in [17,60]. The most efficient, with $7n$ memory locations (additional memory for carrying out the exchanges included) is evidently that of Renka([17]). The method of Lawson([60]) requires about $18n$, and the subroutine IDTANG([5]) of Akima([4,5,17]), $32n$. As far as speed goes, Lawson's method (for which corresponding software has been described but not published) is the fastest and Akima's the slowest. Renka's method with presorting of the data is not much slower than that of Lawson. On a large computer, the times for all three methods applied to about 1,000 points are on the order of a second! The subroutine package ([84]) of Renka uses the least memory at the cost of a relatively small increase in execution time compared to other methods ([17]). Moreover, it is organized in such a way that, among other things, other nodes may easily be added if necessary.

The user-friendly software package of Renka([84]) has altogether more than 100 pages of FORTRAN statements, although many of these are comment statements. About half the total 35 subroutines are used for calculating the triangulation. The user almost always manages by calling REORDR (to sort the nodes) and TRMESH (construction of an optimal triangulation). If a plot is desired, then in addition TRPLOT would have to be adapted to the available plotter software. The subroutines are very well-documented. Their mathematical details and execution time comparisons on an IBM 3033 may be found in [17].

Figure 7.2 shows one so-constructed optimal triangulation for $n = 200$ points chosen at random from a certain rectangle. There are $N = 384$ triangles, $K = 583$ edges and $n_R = 14$ boundary points. For this problem, TRMESH needed .516 seconds of cpu time on a Siemens 7890C computer. For other examples of $n = 36$, 49, 75, and 500 points, the corresponding times were .021, .037, .095, and 2.707 seconds.

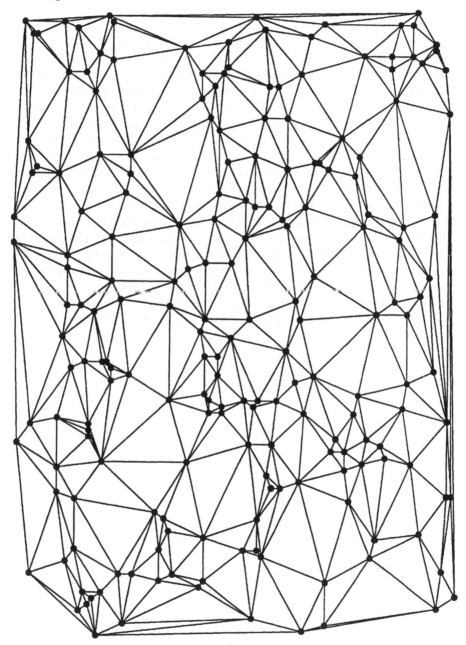

Figure 7.2.

The previously described criterium for optimal triangulations (with possibly equal triangle angles) depends only on the set P of nodes in the xy-plane; the information on the values to be interpolated, z_i, $i = 1, \cdots, n$, are (just as for other criteria) not used. Hence, lately, people have been looking for *data-dependent criteria*, meaning the dependence of the triangulation of x, y, *and* z values. In [27], various measures of the discontinuity of the normal vectors to C^0 linear spline interpolants (planar segments joined continuously) over triangles (see the next chapter) are considered and correspondingly optimal triangulations determined. If the data z_i come from a function with some special properties, then this results in a triangulation with very many long, skinny triangles that quite surprisingly gives a visually smoother-appearing interpolation function than does our criterium discussed earlier. Similarly, although not quite as succesfully, this is also the case when one uses C^1 bicubic spline interpolation (see Chapter 11) ([79,80]): there are dramatic improvements but only for data that comes from very special functions. If the z_i are arbitrary, measured data, which may be taken to be the usual case, then these methods are no better than our previously described criterium. For this reason, we will base the developments of subsequent chapters on the software package of Renka.

8

Linear Spline Interpolants over Triangulations

As is well-known, three distinct noncollinear points in $\mathbb{R}^3$ uniquely define a plane. If we are given a triangulation T of $P = \{P_1, \cdots, P_n\}$, then such an interpolating plane can be constructed above each of the N triangles. Since the planes above any two neighboring triangles always have a line segment in common, we end up in this way with a C^0 spline interpolant. In contrast to subsequent chapters, we give here the relatively short formulas involved and also introduce barycentric coordinates on a triangle.

Without loss of generality, we may suppose that the three points are the first three points of P, i.e., P_1, P_2, and P_3. These have coordinates (x_i, y_i) with heights z_i to be interpolated, $i = 1, 2, 3$. If we write the interpolating plane in the form,

$$E(x, y) = a + b(x - x_1) + c(y - y_1), \tag{8.1}$$

then the interpolation conditions may be written in matrix form as

$$\begin{pmatrix} 1 & 0 & 0 \\ 0 & x_2 - x_1 & y_2 - y_1 \\ 0 & x_3 - x_1 & y_3 - y_1 \end{pmatrix} \begin{pmatrix} a \\ b \\ c \end{pmatrix} = \begin{pmatrix} z_1 \\ z_2 \\ z_3 \end{pmatrix}. \tag{8.2}$$

Then, with

$$d = (x_2 - x_1)(y_3 - y_1) - (x_3 - x_1)(y_2 - y_1) \neq 0, \tag{8.3}$$

the desired coefficients are given by

$$a = z_1,$$

$$b = \frac{1}{d}[(z_2 - z_1)(y_3 - y_1) - (z_3 - z_1)(y_2 - y_1)], \qquad (8.4)$$

$$c = \frac{1}{d}[(z_3 - z_1)(x_2 - x_1) - (z_2 - z_1)(x_3 - x_1)].$$

The *barycentric coordinates* η_1, η_2, η_3 of a point $Q = (x, y)$ with respect to the triangle Δ with vertices P_1, P_2, and P_3 are defined by the normalization condition,

$$\eta_1 + \eta_2 + \eta_3 = 1, \qquad (8.5)$$

and by the requirement that (Q a weighted linear combination of P_1, P_2, P_3)

$$\eta_1 \begin{pmatrix} x_1 \\ y_1 \end{pmatrix} + \eta_2 \begin{pmatrix} x_2 \\ y_2 \end{pmatrix} + \eta_3 \begin{pmatrix} x_3 \\ y_3 \end{pmatrix} = \begin{pmatrix} x \\ y \end{pmatrix}. \qquad (8.6)$$

With the notation,

$$u_i = x_{i+2} - x_{i+1}, \quad v_i = y_{i+2} - y_{i+1}, \qquad i = 1, 2, 3, \qquad (8.7)$$

where the indices on the right are to be taken modulo three, then the linear system of equations (8.5) and (8.6) may be solved for

$$\eta_i = \eta_i(x, y) = \frac{1}{d}[u_i(y - y_{i+1}) - v_i(x - x_{i+1})], \qquad (8.8)$$

with d as in (8.3).

The linear functions of x and y, η_i, have the following properties:

$$\begin{aligned} \eta_i(P_j) &= & \delta_{ij} & \quad i, j = 1, 2, 3, \\ \eta_i(Q) = 0 &\iff & Q \in \overline{P_{i+1}P_{i+2}} & \quad i = 1, 2, 3 \\ \eta_i(Q) \geq 0 \text{ for } i = 1, 2, 3 &\iff & Q \in \Delta, & \qquad (8.9) \\ \eta_i(Q) < 0 \text{ for some } i &\iff & Q \notin \Delta. \end{aligned}$$

By the linearity of (8.1) and (8.6), we have, for $Q \in \Delta$, the interpolation formula,

$$\begin{aligned} E(x, y) = E(Q) &= & \eta_1 E(P_1) + \eta_2 E(P_2) + \eta_3 E(P_3) \\ &= & \eta_1 z_1 + \eta_2 z_2 + \eta_3 z_3. \qquad (8.10) \end{aligned}$$

This method is implemented in the subroutine INTRC0 of the sofware package [84] that was partly introduced in the previous chapter. INTRC0 uses the subroutines TRFIND (to find that triangle in the Delaunay triangulation of P that contains a given point Q of the convex hull of P)

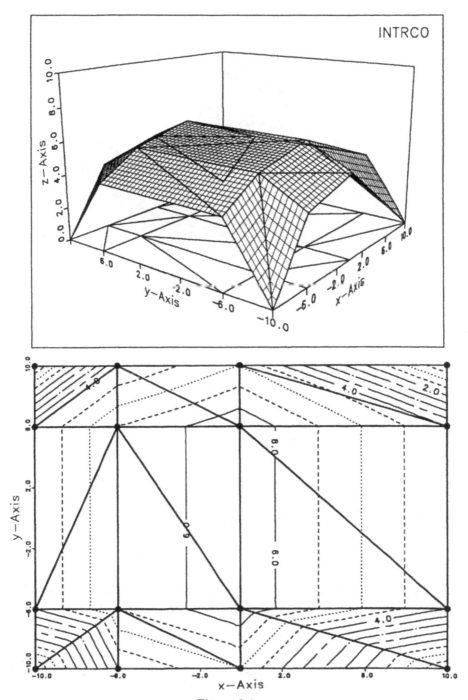

Figure 8.1.

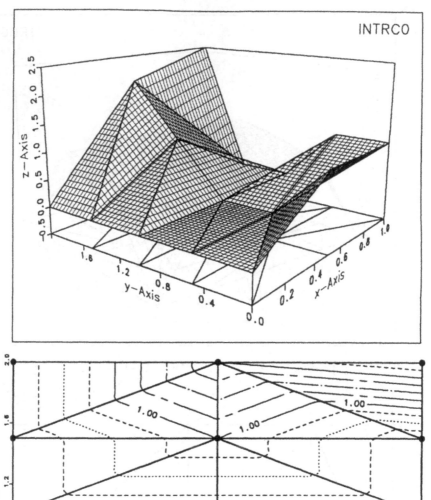

Figure 8.2.

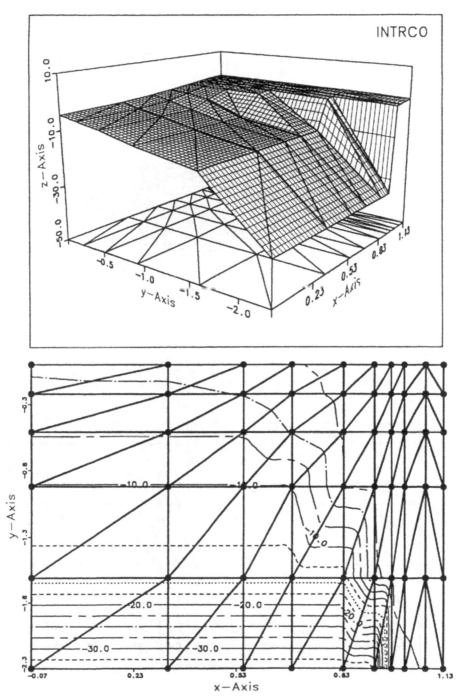

Figure 8.3.

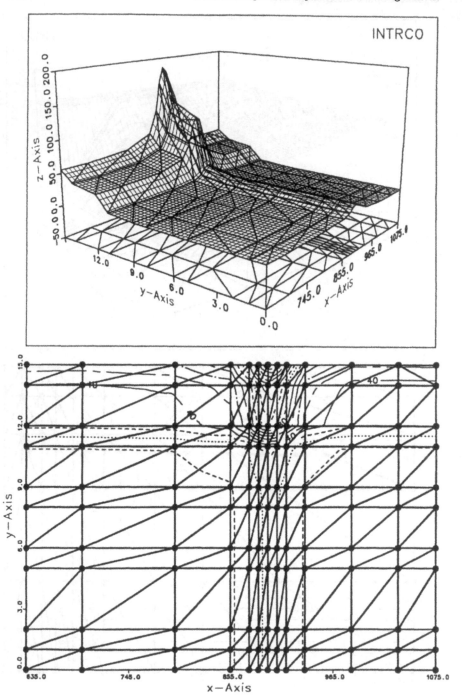

Figure 8.4.

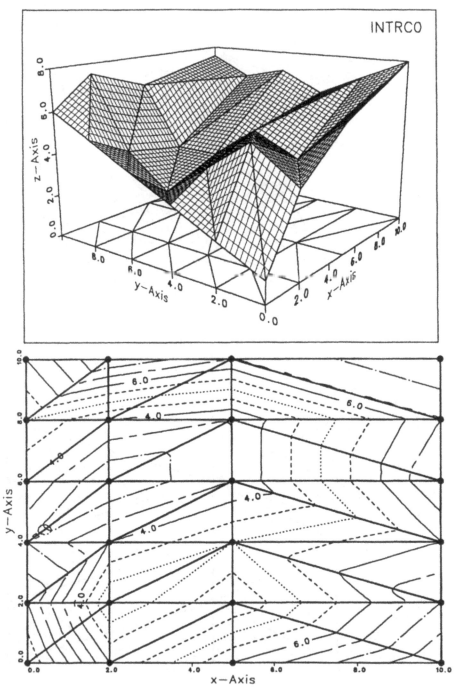

Figure 8.5.

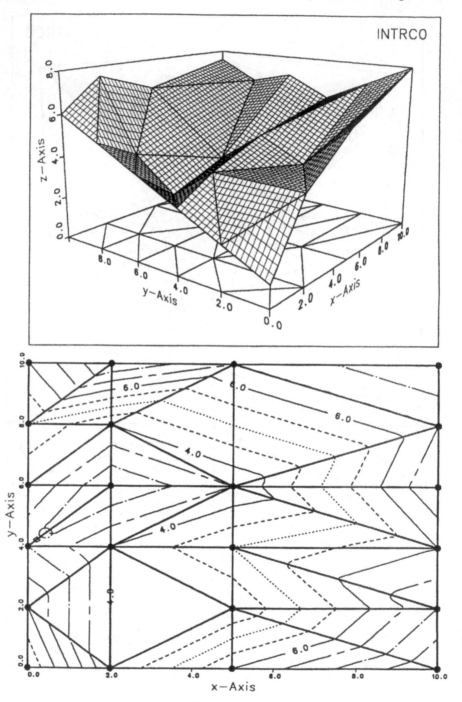

Figure 8.6.

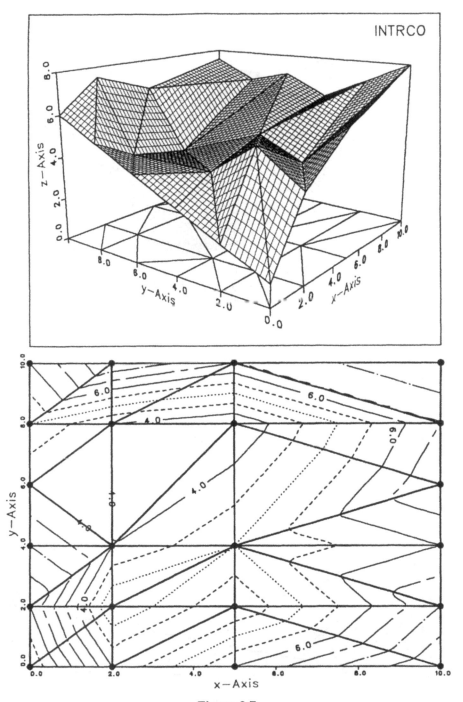

Figure 8.7.

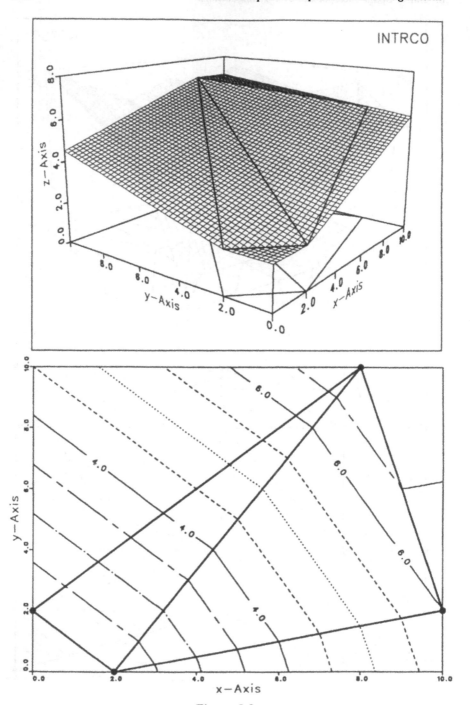

Figure 8.8.

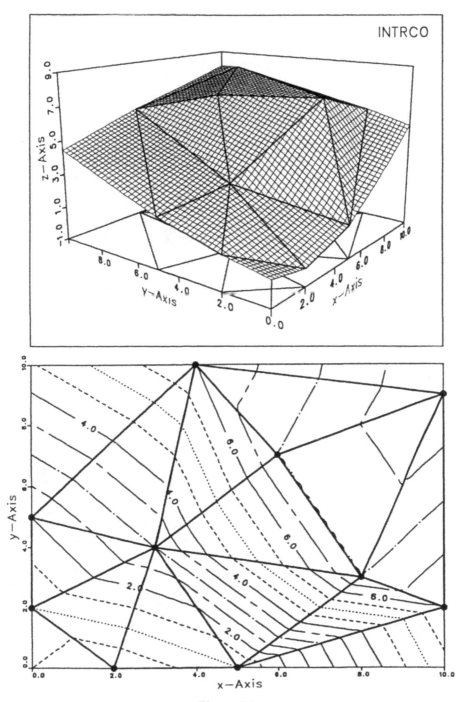

Figure 8.9.

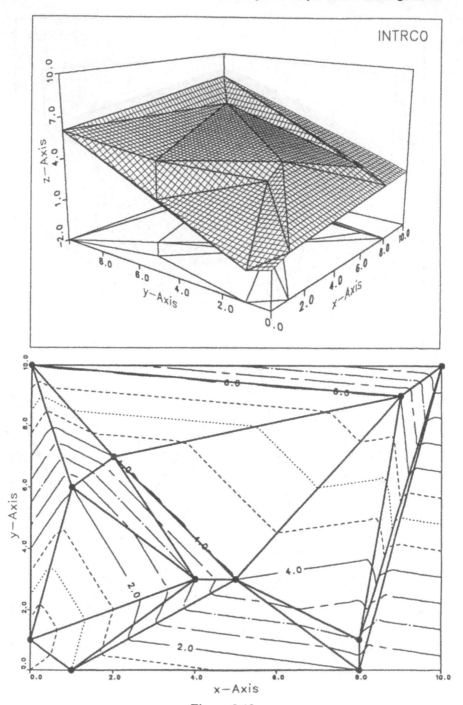

Figure 8.10.

and COORDS (calculation of the nonnegative barycentric coordinates of Q with respect to the triangle just found). Evaluation is then done according to (8.10). In the case that Q is not within the convex hull of P, INTRC0 extends the C^0 spline interpolant continuously to the outside: $E(Q)$ is set to $E(R)$, where R is the point on the boundary nearest to Q. This extrapolation allows the interpolating surface to be extended continuously outside the convex hull of P to, for example, the smallest enclosing rectangle with sides parallel to the axes. This is necessary in many situations; for example, in plotting.

In the first five examples, calculated with INTRC0, shown in Figs. 8.1–8.5, the data were simply taken from Part I, i.e., the P_i, $i = 1, \cdots, n$, lie on a rectangular grid. The resulting Delaunay triangulations are shown with heavy lines in the lower halves of the figures, but also, as much as possible, in the upper halves. Moreover, they are projected onto the interpolating surfaces. These should be compared with those of C^0 bilinear interpolation and C^2 bicubic spline interpolation (Figs. 2.4, 2.7, 2.11, 2.13, 2.5 and 4.9, 4.10, 4.15, 4.17, 4.8, respectively). The examples in Figs. 8.6 and 8.7 were obtained from Fig. 8.5 by removing some of the interior points of the rectangular grid. The convex hull of P is, in this case, a rectangle. Figures 8.8–8.10 show the effect of extrapolation. Figures 8.9 and 8.10 can be compared with Figs. 6.25 and 6.26. We will make use of these examples in subsequent chapters.

We would like once more to mention data-dependent C^0 spline interpolation ([27]), which for very special point configurations can lead to interpolating surfaces better than those based on Delaunay triangulations.

9

The Approximation of First Partial Derivatives

Estimates of the first partial derivatives at the P_i, $i = 1, \cdots, n$, will be needed in Chapters 10 and 11. In Chapter 12, second derivatives will also be used; we will not go into this until then. We need to make these estimates, since the number of polynomial coefficients per triangle being sought is always larger than the (three) interpolation conditions at the vertices. If one desires to treat the three vertices of a triangle symmetrically, then even a subdivision (Chapters 10 and 11) into subtriangles becomes necessary in order to bring the number of polynomial coefficients and the number of interpolation conditions (z_i values and derivatives) into agreement.

Approximation of partial derivatives was easy for the rectangular grids of Part I: one could use well-known univariate methods along each grid line. This, of course, cannot be done for irregular triangulations.

One obvious approach is to use locally the global interpolation methods already described in Part II and to analytically differentiate and evaluate the resulting function at P_i. Thus, it is suggested in [103] to use Shepard's method in the following manner.

For each point P_i, the $S_0(x, y)$ of (6.6) is determined for the NCP closest points to P_i and then to calculate

$$zx_i = \left.\frac{\partial S_0}{\partial x}\right|_{(x_i,y_i)}, \qquad zy_i = \left.\frac{\partial S_0}{\partial y}\right|_{(x_i,y_i)}. \qquad (9.1)$$

NCP=6 is suggested. It is essential that the point P_i itself not be involved in S_0, since otherwise the desired values of zx_i and zy_i would be zero. A very good version of such a method may be found in [85].

Hardy's method can be used in an analogous manner. Thus, it is suggested in [103] to calculate $B(r_k)$ of (6.21) with $q = 1/2$ for P_i itself and its NCP=19 closest neighbors and then to determine zx_i and zy_i by differ-

```
      SUBROUTINE GRADH(NDIM,N,X,Y,Z,Q,R,NCP,ZXZY,
     &                 IPC,IWORK,XX,YY,ZZ,WORK,A,B,IER)
      INTEGER NDIM,N,NCP,IPC(NCP*N),IWORK(N),IER
      REAL X(N),Y(N),Z(N),Q,R,ZXZY(2,N),
     &     XX(N),YY(N),ZZ(N),WORK(N),A(N),B(NDIM,N)
      IER=0
      CALL CLDP(NDIM,N,X,Y,IPC,IWORK,WORK,IER)
      IF (IER.LT.0) RETURN
      IR=0
      IF (R.LE.0.0) IR=1
      DO 30 K=1,N
         XX(1)=X(K)
         YY(1)=Y(K)
         ZZ(1)=Z(K)
         DO 10 I=1,NCP
            I1=IPC((K-1)*NCP+I)
            XX(I+1)=X(I1)
            YY(I+1)=Y(I1)
            ZZ(I+1)=Z(I1)
10       CONTINUE
         CALL HARDY(NDIM,NCP+1,Q,R,XX,YY,ZZ,A,B,IWORK,COND,WORK)
         R2=R*R
         ZX=0.0
         ZY=0.0
         XX1=XX(1)
         YY1=YY(1)
         DO 20 I=2,NCP+1
            DX=XX1-XX(I)
            DY=YY1-YY(I)
            SXY=A(I)*(DX*DX+DY*DY+R2)**(Q-1.0)
            ZX=ZX+SXY*DX
            ZY=ZY+SXY*DY
20       CONTINUE
         ZXZY(1,K)=2.0*Q*ZX
         ZXZY(2,K)=2.0*Q*ZY
         IF (IR.EQ.1) R=-1.0
30    CONTINUE
      RETURN
      END
```

Figure 9.1. Program listing of GRADH.

Calling sequence:

CALL GRADH(NDIM,N,X,Y,Z,Q,R,NCP,ZXZY,
          IPC,IWORK,W1,W2,W3,W4,W5,B,IER)

Purpose:

For points (x_i, y_i) $i = 1, \cdots, n$, arbitrarily distributed in the plane, and associated heights z_i, estimates for the gradients at the nodes are calculated. To this purpose, for each node (x_k, y_k), $k = 1, \cdots, n$, the Hardy multiquadric interpolant of it and its NCP nearest neighbors is computed. Then the gradient of the Hardy surface is computed at the point (x_k, y_k).

Description of the parameters:

NDIM	Maximum first dimension of B.
N	Number of given points.
X	ARRAY(N): x-coordinates of the nodes.
Y	ARRAY(N): y-coordinates of the nodes.
Z	ARRAY(N): Heights z_i, $i = 1, \cdots, n$.
Q	Value of the exponent q used in Hardy's method.
R	Value of the quantity R used in Hardy's method. If $R \leq 0.0$ is given, then R is set to SQRT(AMAX1(XD,YD)/10.) (with one decimal place). Here, XD is the difference between the largest and smallest x values. YD is analogous for the y values.
NCP	Number of neighboring nodes (which will be locally determined).
ZXZY	ARRAY(2,N): Output: Gradient estimates.
	ZXZY(1,K): Partial with respect to x at the kth node.
	ZXZY(2,K): Partial with respect to y at the kth node.
IPC	ARRAY(NCP∗N): Work space.
IWORK, W1,W2,W3,W4,W5	ARRAY(N): Work space.
B	ARRAY(NDIM,N): Work space.
IER	=0: Normal execution.
	< 0: Error in CLDP.

Required subroutines: CLDP, HARDY.

Figure 9.2. Description of GRADH.

entiation. This, for example, results in

$$zx_i = \left.\frac{\partial H}{\partial x}\right|_{(x_i, y_i)} = 2q \sum_{k \neq i} a_k (r_k^2 + R^2)^{q-1}(x_i - x_k). \qquad (9.2)$$

This is implemented in the subroutine GRADH (Figs. 9.1 and 9.2) by means of HARDY and CLDP (Figs. 9.3 and 9.4). CLDP comes from [68] and is based on IDCLDP of [5]. For each $P_i = (x_i, y_i)$ $i = 1, \cdots, n$, it finds the NCP nearest neighbors of P_i. Here, NCP $= \min(n - 1, 19)$ is a pretty reasonable choice. We will return to GRADH in Chapter 11.

In the next techniques([70]), planes E_{ijk} are first passed through P_i and all pairs of points (P_j, P_k) that are in a certain neighborhood of P_i $(j, k \in N_i$ $j \neq k$, $|N_i| \geq 2)$. These have the form,

$$E_{ijk}(x, y) = z_i + b_i(i, j, k)(x - x_i) + c_i(i, j, k)(y - y_i). \qquad (9.3)$$

```
      SUBROUTINE CLDP(NDIM,N,X,Y,NCP,IPC,IPCO,DSQO,IER)
      INTEGER NDIM,N,NCP,IPC(NCP*N),IPCO(NCP),IER
      REAL X(N),Y(N),DSQO(NCP)
      IER=0
      IF (N.LT.2.OR.N.GT.NDIM) THEN
          IER=-1
          RETURN
      END IF
      IF (NCP.LT.1.OR.NCP.GE.N) THEN
          IER=-2
          RETURN
      END IF
      DO 130 IP1=1,N
          X1=X(IP1)
          Y1=Y(IP1)
          J1=0
          DSQMX=0.0
          DO 20 IP2=1,N
              IF (IP2.EQ.IP1) GOTO 20
              DELTAX=X(IP2)-X1
              DELTAY=Y(IP2)-Y1
              DSQI=DELTAX*DELTAX+DELTAY*DELTAY
              J1=J1+1
              DSQO(J1)=DSQI
              IPCO(J1)=IP2
              IF (DSQI.LE.DSQMX) GOTO 10
              DSQMX=DSQI
              JMX=J1
10            IF (J1.GE.NCP) GOTO 30
20        CONTINUE
30        IP2MN=IP2+1
          IF (IP2MN.GT.N) GOTO 60
          DO 50 IP2=IP2MN,N
              IF (IP2.EQ.IP1) GOTO 50
```

(cont.)

```
                DELTAX=X(IP2)-X1
                DELTAY=Y(IP2)-Y1
                DSQI=DELTAX*DELTAX+DELTAY*DELTAY
                IF (DSQI.GE.DSQMX) GOTO 50
                DSQ0(JMX)=DSQI
                IPC0(JMX)=IP2
                DSQMX=0.0
                DO 40 J1=1,NCP
                    IF (DSQ0(J1).LE.DSQMX) GOTO 40
                    DSQMX=DSQ0(J1)
                    JMX=J1
40              CONTINUE
50      CONTINUE
60      IP2=IPC0(1)
        DX12=X(IP2)-X1
        DY12=Y(IP2)-Y1
        DO 70 J3=2,NCP
            IP3=IPC0(J3)
            DX13=X(IP3)-X1
            DY13=Y(IP3)-Y1
            IF ((DY13*DX12-DX13*DY12).NE.0.0) GOTO 110
70      CONTINUE
        NCLPT=0
        DO 100 IP3=1,N
            IF (IP3.EQ.IP1) GOTO 100
            DO 80 J4=1,NCP
                IF (IP3.EQ.IPC0(J4)) GOTO 100
80          CONTINUE
            DX13=X(IP3)-X1
            DY13=Y(IP3)-Y1
            IF ((DY13*DX12-DX13*DY12).EQ.0.0) GOTO 100
            DELTAX=X(IP3)-X1
            DELTAY=Y(IP3)-Y1
            DSQI=DELTAX*DELTAX+DELTAY*DELTAY
            IF (NCLPT.EQ.0) GOTO 90
            IF (DSQI.GE.DSQMN) GOTO 100
90          NCLPT=1
            DSQMN=DSQI
            IP3MN=IP3
100     CONTINUE
        IF (NCLPT.EQ.0) THEN
            IER=-3
            RETURN
        END IF
        DSQMX=DSQMN
        IPC0(JMX)=IP3MN
110     J1=(IP1-1)*NCP
        DO 120 J2=1,NCP
            J1=J1+1
            IPC(J1)=IPC0(J2)
120     CONTINUE
133 CONTINUE
    RETURN
    END
```

Figure 9.3. Program listing of CLDP.

Calling sequence:

CALL CLDP(NDIM,X,Y,NCP,IPC,IWORK,WORK,IER)

Purpose:
This subroutine finds for each node (x_k, y_k), $k = 1, \cdots, n$, its NCP nearest neighbors with respect to Euclidean distance.

Description of the parameters:

NDIM,N,X,Y as in HARDY.

NCP   Number of nearest neighbors that are to be determined for each node. Restriction: $1 \leq$ NCP $<$ N.

IPC   ARRAY(N*NCP): Output: Indices for the nearest neighbors. Node K has neighbors IPC$((K-1)*$NCP$+1), \cdots$, IPC$(K*$NCP$)$, $1 \leq$ K $\leq$ N.

WORK,IWORK   ARRAY(N): Work space.

IER $=0$:   Normal execution.
     $=-1$:   N $<$ 2 or N $>$ NDIM.
     $=-2$:   NCP $<$ 1 or NCP $\geq$ N.
     $=-3$:   All the knots are collinear.

Figure 9.4. Description of CLDP.

The coefficients,

$$b_i(i, j, k) = \frac{\partial E_{ijk}}{\partial x}, \qquad c_i(i, j, k) = \frac{\partial E_{ijk}}{\partial y}, \qquad (9.4)$$

are to be determined so that

$$E_{ijk}(x_j, y_j) = z_j, \qquad E_{ijk}(x_k, y_k) = z_k. \qquad (9.5)$$

Then a convex combination of these planes is taken, i.e., we take the weighted sum,

$$\overline{E}_i(x, y) = \frac{1}{w_i} \sum_{i,j \in N_i} w_{ijk} E_{ijk}(x, y), \qquad (9.6)$$

where

$$w_i = \sum_{j,k \in N_i} w_{ijk} \qquad (9.7)$$

and $w_{ijk} \geq 0$. The desired estimates are then taken as the partial derivatives of the plane $\overline{E}_i$, i.e.,

$$zx_i \;=\; \left.\frac{\partial \overline{E}_i}{\partial x}\right|_{(x_i,y_i)} \;=\; \frac{1}{w_i}\sum_{j,k\in N_i} w_{ijk}b_i(i,j,k), \tag{9.8}$$

$$zy_i \;=\; \left.\frac{\partial \overline{E}_i}{\partial y}\right|_{(x_i,y_i)} \;=\; \frac{1}{w_i}\sum_{j,k\in N_i} w_{ijk}c_i(i,j,k). \tag{9.9}$$

```
      SUBROUTINE GRADLK(NDIM,N,X,Y,Z,IADJ,IEND,ZXZY,IADJTR,
     &                  IENDTR,ITR,IER)
      INTEGER NDIM,N,IADJ(6*N-9),IEND(N),IADJTR(6*N),
     &        IENDTR(N),ITR(6*N),IER
      REAL X(N),Y(N),Z(N),ZXZY(2,N)
      IER=0
      IF (N.LT.3.OR.N.GT.NDIM) THEN
         IER=-1
         RETURN
      END IF
      CALL TRLIST(N,IADJ,IEND,IADJTR,IENDTR,ITR)
      DO 20 I=1,N
         ZXZY(1,I)=0.0
         ZXZY(2,I)=0.0
         IF (I.EQ.1) THEN
            NSTART=1
         ELSE
            NSTART=IENDTR(I-1)+1
         END IF
         NEND=IENDTR(I)
         DO 10 K=NSTART,NEND
            KT=IADJTR(K)
            J=3*KT
            I3=ITR(J)
            I2=ITR(J-1)
            I1=ITR(J-2)
            U2=X(I2)-X(I1)
            U3=X(I3)-X(I1)
            V2=Y(I2)-Y(I1)
            V3=Y(I3)-Y(I1)
            W2=Z(I2)-Z(I1)
            W3=Z(I3)-Z(I1)
            DELTA=U2*V3-V2*U3
            ZXZY(1,I)=ZXZY(1,I)+(W2*V3-V2*W3)/DELTA
            ZXZY(2,I)=ZXZY(2,I)+(U2*W3-W2*U3)/DELTA
   10    CONTINUE
         M=NEND-NSTART+1
         ZXZY(1,I)=ZXZY(1,I)/FLOAT(M)
         ZXZY(2,I)=ZXZY(2,I)/FLOAT(M)
   20 CONTINUE
      RETURN
      END
```

Figure 9.5. Program listing of GRADLK.

Calling sequence:

CALL GRADLK(NDIM,N,X,Y,Z,IADJ,IEND,ZXZY,I1,I2,I3,IER)

Purpose:
For given nodes (x_k, y_k), $k = 1, \cdots, n$, and associated heights z_k, approximate values for the first partial derivatives at these points are calculated by a method due to Klucewicz.

Description of the parameters:

NDIM,N,X,Y,Z as in HARDY.
IADJ   ARRAY(6 * N − 9).
IEND   ARRAY(N).
             IADJ and IEND are used for keeping track of the Delaunay triangulation. This triangulation is computed by a call, separate from GRADLK, to the subroutine TRMESH of ACM624.
ZXZY   ARRAY(2,N): Output: Partial derivatives at the knots.
             ZXZY(1,K): Partial with respect to x at the Kth node.
             ZXZY(1,K): Partial with respect to y at the Kth node.
I1   ARRAY(6*N): Work space.
I2   ARRAY(N): Work space.
I3   ARRAY(6*N): Work space.
IER   =0:   Normal execution.
         =−1:   N < 3 or N > NDIM.

Required subroutines: TRLIST, INSERT.

Figure 9.6. Description of GRADLK.

```
      SUBROUTINE TRLIST(N,IADJ,IEND,IADJTR,IENDTR,ITR)
      INTEGER N,IADJ(6*N-9),IEND(N),IADJTR(6*N),IENDTR(N),
     &          ITR(6*N)
      LOGICAL BNODE
      INDXTR=0
      INDEX=0
      DO 20 IK=1,N
          BNODE=.FALSE.
          IF (IK.EQ.1) THEN
              NSTART=1
          ELSE
              NSTART=IEND(IK-1)+1
          END IF
          NEND=IEND(IK)
```

(cont.)

```
            IF (IADJ(NEND).EQ.0) THEN
                NEND=NEND-1
                BNODE=.TRUE.
            END IF
            I1=IK
            DO 10 I=NSTART,NEND-1
                I2=IADJ(I)
                I3=ADJ(I+1)
                CALL INSERT(I1,I2,I3,INDXTR,INDEX,IADJTR,ITR)
 10         CONTINUE
            IF (.NOT.BNODE) THEN
                I2=IADJ(NEND)
                I3=IADJ(NSTART)
                CALL INSERT(I1,I2,I3,INDXTR,INDEX,IADJTR,ITR)
            END IF
            IENDTR(IK)=INDEX
 20     CONTINUE
        RETURN
        END
```

Figure 9.7. Program listing of TRLIST.

```
        SUBROUTINE INSERT(I1,I2,I3,INDXTR,INDEX,IADJTR,ITR)
        INTEGER I1,I2,I3,INDXTR,INDEX,IADJTR(1),ITR(1)
        IF (I2.GT.I1.AND.I3.GT.I1) THEN
            INDXTR=INDXTR+1
            J3=3*INDXTR
            ITR(J3-2)=I1
            ITR(J3-1)=I2
            ITR(J3)=I3
            INDEX=INDEX+1
            IADJTR(INDEX)=INDXTR
        ELSE
            DO 10 KTR=1,NDXTR
                J3=3*KTR
                J2=J3-1
                J1=J3-2
                IF ( ((ITR(J1).EQ.I1).AND.(ITR(J2).EQ.I2).AND.
     &               (ITR(J3).EQ.I3)).OR.
     &               ((ITR(J1).EQ.I2).AND.(ITR(J2).EQ.I3).AND.
     &               (ITR(J3).EQ.I1)).OR.
     &               ((ITR(J1).EQ.I3).AND.(ITR(J2).EQ.I1).AND.
     &               (ITR(J3).EQ.I2))) THEN
                    INDEX=INDEX+1
                    IADJTR(INDEX)=KTR
                END IF
 10         CONTINUE
        END IF
        RETURN
        END
```

Figure 9.8. Program listing of INSERT.

Special methods are so obtained through the choice of the weights and the index sets N_i.

Klucewicz's method([57]) results when the weights are taken to be $w_{ijk} = 1$ and the index set N_i as the set of indices of the vertices of those triangles for which P_i is also a vertex. This method is implemented in the subroutine GRADLK([68]) using the data structure of [85]. Hence, GRADLK assumes a triangulation computed by TRMESH (see Chapter 8). Figure 9.5 contains the listing and 9.6 the program description. It calls the subroutines TRLIST (Fig. 9.7) (to find the relevant triangles) and INSERT (Fig. 9.8, [68]).

Akima's method ([4,5]) is also of this type([70]). All the cross products of the vectors, $\overrightarrow{P_iP_j}$ and $\overrightarrow{P_iP_k}$, $j, k \in N_i$, $j \neq k$, from the NCP nearest neighbors of P_i are formed. These are standardized so that the third components are nonzero and then averaged. The plane, normal to the resulting vector (α, β, γ) (with $\gamma > 0$), has equation,

$$\alpha x + \beta y + \gamma z + \delta = 0. \tag{9.10}$$

This may be solved for z to give

$$\overline{E}(x, y) = z = -\frac{\alpha}{\gamma}x - \frac{\beta}{\gamma}y - \frac{\delta}{\gamma}, \tag{9.11}$$

which results in the estimates,

$$zx_i = -\frac{\alpha}{\gamma}, \qquad zy_i = -\frac{\beta}{\gamma}. \tag{9.12}$$

Naturally, the method (subroutine IDPDRV) is among the programs [5]. We give here a subroutine GRADLA (Figs. 9.9 and 9.10, [68]) that uses the subroutine CLDP (which we have already introduced) to determine the NCP nearest neighbors. The modifications of the method suggested by Akima himself in [6], where in particular the user no longer has to set the value of NCP, do not always seem to be an improvement. Nevertheless, the IMSL subroutine SURF does incorporate them (see also Chapter 12).

The most successful of this class of methods ([70]) is probably that of Little([62]), where the N_i are chosen as in Akima's method (NCP=$n-1$ permitted) and the weights set to

$$w_{ijk} = r_{ij}^{-2}r_{ik}^{-2}, \tag{9.13}$$

where r_{ij} and r_{ik} are the Euclidean lengths of the segments $\overline{P_iP_j}$ and $\overline{P_iP_k}$, respectively. It would be a simple matter to give an implementation based on GRADLA, but we will not do so.

In Lawson's method([60]), for each point P_i a quadratic polynomial of type (6.3) that interpolates at P_i, i.e., is of the form,

$$q(x,y) = z_i + zx_i(x-x_i) + zy_i(y-y_i) + a(x-x_i)^2 + b(x-x_i)(y-y_i) + c(y-y_i)^2,$$
(9.14)

is fitted in the sense of weighted least squares at the vertices of all the immediately neighboring triangles (as in Klucewicz).

Renka ([81,85]) chooses these weights according to

$$w_k = \begin{cases} \dfrac{1}{d_k} - \dfrac{1}{r_i} & \text{if } r_i > d_k \\ 0 & \text{otherwise} \end{cases}.$$
(9.15)

Here, d_k is the distance from P_k to P_i. If D is the distance from P_i to that point P_j that is farthest away if $n < 9$ and is eighth nearest if $n \geq 9$, then the radius of influence r_i in (9.15) is defined as the distance between P_i and the nearest point P_ℓ whose distance to P_i is larger than D. If there is no such point P_ℓ, then r_i is set to $r_i = 2D$. If $n < 6$, then a plane is used instead of a quadratic. For $n \geq 6$, care is taken to ensure the solvability of the resulting normal equations. This method is implemented in the subroutine GRADL of Renka's software package [85].

There also is available another subroutine, GRADG, that is based on an idea of Nielson([69]). It is global in the sense that it uses all the given points. We refer to [69] for the details. GRADG should be about five

```
      SUBROUTINE GRADLA(NDIM,N,X,Y,Z,NCP,ZXZY,IPC,IWORK,
     &                  WORK,IER)
      INTEGER NDIM,N,NCP,IER,IPC(NCP*N),IWORK(NCP)
      REAL X(N),Y(N),Z(N),ZXZY(2,N),WORK(NCP)
      REAL NMX,NMY,NMZ,NMXX,NMXY,NMYX,NMYY
      IER=0
      CALL CLDP(NDIM,N,X,Y,NCP,IPC,IWORK,WORK,IER)
      IF (IER.NE.0) RETURN
      NCPM1=NCP-1
      DO 40 IPO=1,N
         XO=X(IPO)
         YO=Y(IPO)
         ZO=Z(IPO)
         NMX=0.0
         NMY=0.0
         NMZ=0.0
         JIPCO=NCP*(IPO-1)
         DO 30 IC1=1,NCPM1
            JIPC=JIPCO+IC1
            IPI=IPC(JIPC)
            DX1=X(IPI)-XO
            DY1=Y(IPI)-YO
            DZ1=Z(IPI)-ZO
            IC2MN=IC1+1
```

(*cont.*)

```
            DO 20 IC2=IC2MN,NCP
            JIPC=JIPC0+IC2
            IPI=IPC(JIPC)
            DX2=X(IPI)-X0
            DY2=Y(IPI)-Y0
            DNMZ=DX1*DY2-DY1*DX2
            IF (DNMZ.EQ.0.0) GOTO 20
            DZ2=Z(IPI)-Z0
            DNMX=DY1*DZ2-DZ1*DY2
            DNMY=DZ1*DX2-DX1*DZ2
            IF (DNMZ.GE.0.0) GOTO 10
            DNMX=-DNMX
            DNMY=-DNMY
            DNMZ=-DNMZ
10          NMX=NMX+DNMX
            NMY=NMY+DNMY
            NMZ=NMZ+DNMZ
20          CONTINUE
30      CONTINUE
        ZXZY(1,IP0)=-NMX/NMZ
        ZXZY(2,IP0)=-NMY/NMZ
40  CONTINUE
    RETURN
    END
```

Figure 9.9. Program listing of GRADLA.

Calling sequence:

CALL GRADLA(NDIM,N,X,Y,Z,NCP,ZXZY,IPC,I1,W1,IER)

Purpose:
For given nodes (x_k, y_k), $k = 1, \cdots, n$, and associated heights z_k, approximate values for the first partial derivatives at these nodes are calculated by a method due to Akima.

Description of the parameters:

NDIM,N,X,Y,Z as in HARDY.
NCP as in CLDP.
ZXZY as in GRADLK.

IER	=0:	No error detected.
	< 0:	Error in CLDP.
IPC	ARRAY(N*N): Work space.	
I1	ARRAY(N): Work space.	
W1	ARRAY(N): Work space.	

Required subroutine: CLDP.

Figure 9.10. Description of GRADLA.

times faster than GRADL (for all points). In our experience, which we will explain more fully in Chapter 11, which of the two methods results in a visually better interpolating surface depends on the data.

In [67], a subroutine PDMIN is given that attempts to improve upon Nielson's method [69]. It calculates an initial approximation using the method of Klucewicz in a subroutine PDSTE. We will return to this in Chapter 11.

We will compare several of the aforementioned methods on concrete data in the next chapter. A ranking that holds in general cannot be expected, since obviously for each method there is some dataset for which it does the best.

10

Quadratic Spline Interpolants over Triangulations

In going from C^0 linear spline interpolation, where three coefficients are to be determined per triangle, in agreement with the number of interpolation conditions at the vertices, to the next higher differentiability requirements and hence necessarily to higher-degree polynomials, there arise for triangular meshes unexpected difficulties. Namely, the number of coefficients in (6.3), i.e, 6, 10, 15, and 21 for $n = 2, \cdots, 5$, does not agree with the number of interpolation requirements consistent with the symmetry of the situation. For example, if we have estimates for the first partial derivatives (by symmetry in *both* x and y) at all three vertices (again for reasons of symmetry), then we have $3 + 2 \cdot 3 = 9$ interpolation conditions. If we in addition also have estimates for the second partials, then we would have have $9 + 3 \cdot 3 = 18$ conditions. One way out is always to put additional restrictions on the polynomials (see Chapter 12 for $m = 5$), and another is to subdivide the triangles into subtriangles (in this chapter for $m = 2$ and in Chapter 11 for $m = 3$).

There are many ways to symmetrically split a triangle into subtriangles. The line segments from the centroid S to the three vertices divide the triangle into three subtriangles with equal areas. This will be used in the next chapter for $m = 3$. For $m = 2$, a symmetric subdivision into four subtriangles (by connecting the midpoints of the sides H_1, H_2, and H_3)

does not always suffice for interpolation of the z_i as well as of estimates of the first partials ([78,68]). In contrast, the symmetric divisions into six and 12 subtriangles, as in Fig. 10.1, have proven to be quite useful (U is the circumcenter). The reasons for considering such numbers for a subdivision will become clear once the following assertion has been proved. At the same time, it gives an elegant method for the construction of a C^1 quadratic spline interpolant on a subdivided triangle.

We consider two arbitrary triangles with a common side K. Suppose that this has the linear equation,

$$K: \quad \alpha x + \beta y + \gamma = 0. \tag{10.1}$$

Let q_1 and q_2 be two quadratic functions on the two triangles. Then the global piecewise quadratic function, q, which agrees with q_1 on one triangle and q_2 on the other, is continuously differentiable across the side K if and only if there is a λ such that

$$q_2(x,y) = q_1(x,y) + \lambda(\alpha x + \beta y + \gamma)^2. \tag{10.2}$$

No proof for this is given in [78] but one is available in [68]. Now if (10.2) holds, then by (10.1) it follows immediately that

$$
\begin{aligned}
q_2(x,y) &= q_1(x,y) \quad \text{for all } (x,y) \in K, & (10.3)\\
\nabla q_2(x,y) &= \nabla q_1(x,y) \quad \text{for all } (x,y) \in K. & (10.4)
\end{aligned}
$$

Conversely, suppose that (10.3) and (10.4) hold. In Eq. (10.1) for K, $\alpha = \beta = 0$ is not permitted. Hence, we may assume with out loss that $\alpha \neq 0$. Then we may write

$$
\begin{aligned}
q_1(x,y) = \; & a_1 + a_2(\alpha x + \beta y + \gamma) + a_3 y \\
& + a_4(\alpha x + \beta y + \gamma)^2 + a_5(\alpha x + \beta y + \gamma)y + a_6 y^2
\end{aligned}
$$

and

$$
\begin{aligned}
q_2(x,y) = \; & b_1 + b_2(\alpha x + \beta y + \gamma) + b_3 y \\
& + b_4(\alpha x + \beta y + \gamma)^2 + b_5(\alpha x + \beta y + \gamma)y + b_6 y^2.
\end{aligned}
$$

From (10.3), it follws that for $(x,y) \in K$,

$$a_1 + a_3 y + a_6 y^2 = b_1 + b_3 y + b_6 y^2,$$

and from (10.4) that

$$\alpha a_2 + \alpha a_5 y = \frac{\partial q_1}{\partial x} = \frac{\partial q_2}{\partial x} = \alpha b_2 + \alpha b_5 y.$$

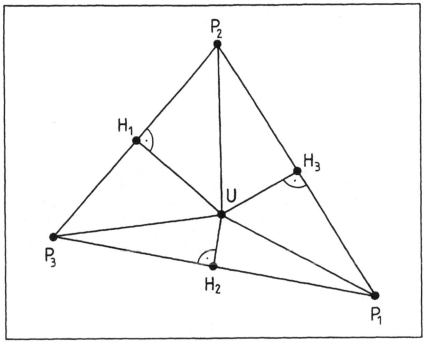

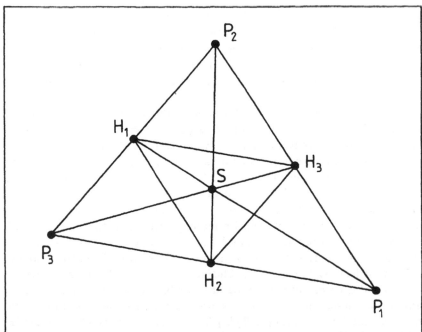

Figure 10.1.

Hence, $a_i = b_i$ for $i = 1, 2, 3, 5, 6$, and so

$$
\begin{aligned}
q_2(x, y) &= a_1 + a_2(\alpha x + \beta y + \gamma) + a_3 y \\
&\quad b_4(\alpha x + \beta y + \gamma)^2 + a_5(\alpha x + \beta y + \gamma) + a_6 y^2 \\
&= q_1(x, y) + (b_4 - a_4)(\alpha x + \beta y + \gamma)^2,
\end{aligned}
$$

i.e., $\lambda = b_4 - a_4$ is uniquely determined.

The construction that now follows of a subdivision such as in Fig. 10.1a does not depend on the choice of U (it can be any point inside the triangle) nor on the choice of H_1, H_2, and H_3 (allowed to be arbitrary points on the three sides). However, because of the ensuing simplifications, we do take the circumcenter (for triangles for which this is inside; other triangles are handled differently) and the midpoints of the sides. We start, for example, in the triangle $P_1 U H_2$ with a quadratic function q_1 having six parameters. From the C^1 conditions across the lines $\overline{P_1 U}$, $\overline{H_3 U}$, $\overline{P_2 U}$, $\overline{H_1 U}$, $\overline{P_3 U}$, and finally $\overline{H_2 U}$, with equations $\alpha_k x + \beta_k y + \gamma_k = 0$, $k = 1, \cdots, 6$, we have the six additional parameters λ_k of (10.2). This gives a total of 12 parameters. Moreover, we must again have

$$
q_6(x, y) + \lambda_6(\alpha_6 x + \beta_6 y + \gamma_6)^2 = q_1(x, y). \tag{10.5}
$$

Without loss, we assume that U is at the origin so that the $\gamma_k = 0$, $k = 1, \cdots, 6$. Thus, there are three conditions imposed by (10.5) that come from comparing the coefficients of x^2, xy, and y^2, and consequently there remain $9 = 12 - 3$ free parameters for the nine interpolation requirements. It is shown in [78,68] that these guarantee the uniqueness of the six quadratic functions.

If all the triangles of a triangulation are subdivided into six, and the six corresponding quadratics determined as above, then the global function Q is continuously differentiable on the whole triangulation ([68,78]). In this assertion, it is now essential that for each the circumcenter is chosen. This, however, is only possible, since U must lie inside the triangle, when all the triangles are acute, which is almost never the case.

For this reason, we now consider a subdivision into 12 subtriangles ([16,52, 53,78]), which can even be more general than that shown in Fig. 10.1b and which in particular may be applied to obtuse triangles. A piecewise quadratic function that is continuously differentiable on the whole triangle has at least 12 parameters. The inner triangle $H_1 H_2 H_3$ is a particular case of a subdivision into six subtriangles which gives nine parameters. In addition to these, another three factors λ arise when we extend as before from the subtriangles $H_1 H_3 S$, $H_1 H_2 S$, and $H_2 H_3 S$ to $H_1 H 3 P 2$, $H 1 H 2 P_3$, and $H_2 H_3 P_1$. If we now also prescribe values for the normal derivatives at H_1, H_2, and H_3, then we have interpolation conditions matching the 12 parameters. In [78,68], it is shown that the desired construction is always possible in this manner. The values for the normal derivatives can be given by linear interpolation, which, for example, gives

$$\frac{\partial q}{\partial n}\bigg|_{H_1} = \frac{1}{2}\left(\operatorname{grad} q|_{P_3} + \operatorname{grad} q|_{P_2}\right) \cdot \mathbf{n}$$

at H_1. This makes the normal derivative linear along the edges, which together with the interpolation conditions again guarantees that the corresponding global function Q is everywhere continuously differentiable on the triangulation.

The two types of subdivision allow a mixed strategy ([78]). If a triangle of a triangulation is sufficiently acute, i.e., if no angle is greater than 75°, then the circumcenter U is likewise sufficiently far inside the triangle and it is subdivided into six subtriangles according to Fig. 10.1 (top). Otherwise — Fig. 10.1 (bottom) is not such a case! — we subdivide into 12 subtriangles. Because of its special properties, subdivision into six triangles is often chosen for a *Delaunay triangulation*. If H_1, H_2, and H_3 are always chosen to be the midpoints of the sides, then despite the two different subdivisions of Figs. 10.1a and b, the C^1 propery of the resulting global function is still guaranteed, since the normal derivatives have been made linear along the edges. In contrast to the original work [78], all the formulas necessary for an implementation of this method are derived in complete detail in [68]. Included are discussions on the application of barycentric coordinates, the choice of a good starting triangle based on symmetry considerations, the generalization of the $m = 2$ case to that of $m = 3$ ([59]), as well as extrapolation outside the triangulation. (Formulas for subdivision into only 12 subtriangles are given in [16].) Since we must, in both of the next two chapters, do without giving all the required formulas — here, the software has been published except without the implemented formulas being available in complete detail, nor easily discernible from the software itself — we refer to the original publication ([68]) and reproduce, with permission of the author, the software therein developed.

The subroutine QUADSF (Figs. 10.2 and 10.3) uses TRFIND and COORDS of [84], assuming in particular that a Delaunay triangulation has been produced by TRMESH([84]) with the data structure used there. QUADSF calls INPOL (Fig. 10.4) for interpolation (TVAL6 (Fig. 10.5) for six and TVAL12 (Fig. 10.6) for 12 subtriangles per triangle) and EXPOL (Fig. 10.7) for extrapolation, as in [81]. HERMI2 (Fig. 10.8) is a routine used by INPOL and EXPOL.

The examples in Figs. 10.9 (cf. Figs. 3.20, 4.8, 5.17), 10.10 (cf. 3.22, 4.10, 4.23, 6.13–6.15, 8.2), 10.11 (cf. 3.23, 4.11, 5.19), and 10.12 (cf. 6.26, 8.10) were all calculated using the gradient estimates determined by GRADL (=GRADLR); GRADLA and GRADLK produced less smooth boundary curves. In Fig. 10.13 the result of interpolation using GRADLK is contrasted with that of using GRADLR (=GRADL) in Fig. 10.12. While only subdivision into 12 subtriangles was used in Figs. 10.9–10.11, because of there being a right angle in each of the triangles, both subdivision possibilities (and some very obtuse triangles) occur in Figs. 10.12 and 10.13.

In general, the results of QUADSF and INTRC1([84]) (see the next chapter) are frequently almost identical. The various gradient routines have a similar influence on both methods. The main advantage of QUADSF could be, as considered by [78], that contour lines are easier to compute for quadratic than for cubic surfaces.

```
      SUBROUTINE QUADSF(N,PX,PY,X,Y,Z,ZXZY,IADJ,IEND,IST,
     &                  PZ,IER)
      INTEGER N,IADJ(1),IEND(1),IST,IER
      REAL PX,PY,X(N),Y(N),Z(N),ZXZY(2,N),PZ
      INTEGER I(3)
      REAL XT(3),YT(3),ZT(3),ZX(3),ZY(3)
      IER=0
      PZ=0.0
      IF (N.GE.3 .AND. IST.GE.1 .AND. IST.LE.N) THEN
          CALL TRFIND(IST,PX,PY,X,Y,IADJ,IEND,I(1),I(2),I(3))
          IF (I(1).NE.0) THEN
              IST=I(1)
              IF (I(3).NE.0) THEN
                  DO 10 J=1,3
                      IJ=I(J)
                      XT(J)=X(IJ)
                      YT(J)=Y(IJ)
                      ZT(J)=Z(IJ)
                      ZX(J)=ZXZY(1,IJ)
                      ZY(J)=ZXZY(2,IJ)
   10             CONTINUE
                  CALL INPOL(PX,PY,XT,YT,ZT,ZX,ZY,PZ,IER)
                  RETURN
              ELSE
                  CALL EXPOL(N,PX,PY,I(1),X,Y,Z,ZXZY,IADJ,IEND,
     &                       PZ,IER)
                  RETURN
              END IF
          ELSE
              IER=-2
              RETURN
          END IF
      ELSE
          IER=-1
          RETURN
      END IF
      END
```

Figure 10.2. Program listing of QUADSF.

Calling sequence:

CALL QUADSF(N,PX,PY,X,Y,Z,ZXZY,IADJ,IEND,IST,PZ,IER)

Purpose:
This subroutine is the driver routine for C^1 quadratic spline interpolation over a triangular grid. The routine TRFIND of ACM624 is used
to find that triangle in which the interpolation point P = (PX,PY)
lies. Then INPOL is called to calculate the interpolant value PZ
at the point P. If P lies outside the convex hull of the given nodes
$\{(x_1, y_1), \cdots, (x_n, y_n)\}$, then an extrapolation is carried out by calling
EXPOL.

Description of the parameters:

N,X,Y,Z as in HARDY.

ZXZY	ARRAY(2,N); Array for the partial derivatives at the nodes:
	ZXZY(1,K): Partial with respect to x at the Kth node
	ZXZY(2,K): Partial with respect to y at the Kth node.
IADJ	ARRAY(6 * N − 9).
IEND	ARRAY(N).
	IADJ and IEND are used for recording the Delaunay
	triangulation of the nodes. This is constructed outside of
	QUADSF by calling the subroutine TRMESH of ACM624.
IST	Input: Node number at which the search is begun for that
	triangle in which P lies. Restriction: $1 \leq$ IST $\leq$ N.
	Output: Number of a node that is a vertex of the last
	triangle processed.
PZ	Calculated function value at the point (PX,PY).
IER	=0: No error, interpolation.
	=1: No error, extrapolation.
	=−1: N or IST is out of allowed bounds.
	=−2: All the nodes as well as P are collinear.

Required subroutines: TRFIND, INPOL, EXPOL, COORDS, TVAL6,
TVAL12, HERMI2.

Figure 10.3. Description of QUADSF.

```
       SUBROUTINE INPOL(PX,PY,X,Y,Z,ZX,ZY,PZ,IER)
       INTEGER IER
       REAL PX,PY,X(3),Y(3),Z(3),ZX(3),ZY(3),PZ
       REAL U(3),V(3),SL(3),THETA(3),ETAS(3),ZP(3),ZXP(3),
     &      ZYP(3)
       IER=0
       PZ=0.0
       U(1)=X(3)-X(2)
       U(2)=X(1)-X(3)
       U(3)=X(2)-X(1)
       V(1)=Y(3)-Y(2)
       V(2)=Y(1)-Y(3)
       V(3)=Y(2)-Y(1)
       SL(1)=SQRT(U(1)*U(1)+V(1)*V(1))
       SL(2)=SQRT(U(2)*U(2)+V(2)*V(2))
       SL(3)=SQRT(U(3)*U(3)+V(3)*V(3))
       DELTA=U(1)*V(2)-U(2)*V(1)
       IF (DELTA.NE.0.0) THEN
          THETA(3)=(-(U(1)*U(2)+V(1)*V(2)))/(SL(1)*SL(2))
          THETA(1)=(-(U(2)*U(3)+V(2)*V(3)))/(SL(2)*SL(3))
          THETA(2)=(-(U(3)*U(1)+V(3)*V(1)))/(SL(3)*SL(1))
          IF (THETA(1).GT.0.26 .AND. THETA(2).GT.0.26
     &                     .AND. THETA(3).GT.0.26) THEN
             C1=(U(1)*(X(3)+X(2))+V(1)*(Y(3)+Y(2)))*0.5
             C2=(U(2)*(X(1)+X(3))+V(2)*(Y(1)+Y(3)))*0.5
             SX=(C1*V(2)-C2*V(1))/DELTA
             SY=(U(1)*C2-U(2)*C1)/DELTA
             CALL COORDS(SX,SY,X(1),X(2),X(3),Y(1),Y(2),Y(3),
     &                     ETAS,IERR)
             IF (IERR.EQ.1) THEN
                IER=-1
                RETURN
             END IF
             CALL TVAL6(PX,PY,X,Y,Z,ZX,ZY,ETAS,PZ,IER)
             RETURN
          ELSE
             CALL TVAL12(PX,PY,X,Y,Z,ZX,ZY,PZ,IER)
             RETURN
          END IF
       ELSE
          IER=-1
          RETURN
       END IF
       END
```

Figure 10.4. Program listing of INPOL.

```
      SUBROUTINE TVAL6(PX,PY,X,Y,Z,ZX,ZY,ETAP,PZ,IER)
      INTEGER IER
      REAL PX,PY,X(3),Y(3),Z(3),ZX(3),ZY(3),ETAP(3),PZ
      REAL SKP
      REAL U(3),V(3),ETA(3),H(3),PHIH(3),PHIK(3),PHIM(3),
     &     DELTA(3)
      SKP(XX,YY,UU,VV)=XX*UU+YY*VV
      IER=0
      PZ=0.0
      U(1)=X(3)-X(2)
      V(1)=Y(3)-Y(2)
      U(2)=X(1)-X(3)
      V(2)=Y(1)-Y(3)
      U(3)=X(2)-X(1)
      V(3)=Y(2)-Y(1)
      DELT=U(1)*V(2)-U(2)*V(1)
      IF (DELT.NE.0.) THEN
      CALL COORDS(PX,PY,X(1),X(2),X(3),Y(1),Y(2),Y(3),
     &               ETA,IERR)
      IF (IERR.EQ.1) THEN
         IER= 1
         RETURN
      END IF
      DETA1=ETAP(3)-ETAP(2)
      DETA2=ETAP(1)-ETAP(3)
      DETA3=ETAP(2)-ETAP(1)
      G1=-ETAP(2)*ETA(1)+DETA2*ETA(2)+ETAP(2)*ETA(3)
      G2=-ETAP(2)*ETA(1)+ETAP(1)*ETA(2)
      G3=DETA1*ETA(1)+ETAP(1)*ETA(2)-ETAP(1)*ETA(3)
      G4=ETAP(3)*ETA(1)-ETAP(1)*ETA(3)
      G5=ETAP(3)*ETA(1)-ETAP(3)*ETA(2)+DETA3*ETA(3)
      G6=-ETAP(3)*ETA(1)+ETAP(2)*ETA(3)
      ALPH13=-0.5*ETAP(2)/ETAP(3)
      ALPH21=-0.5*ETAP(3)/ETAP(1)
      ALPH32=-0.5*ETAP(1)/ETAP(2)
      H(1)=ALPH13*ETA(3)*ETA(3)
      H(2)=ALPH21*ETA(1)*ETA(1)
      H(3)=ALPH32*ETA(2)*ETA(2)
      IF (G6.GT.0. .AND. G1.LE.0.) THEN
         EPSI16=0.5/(ETAP(2)*ETAP(3))
         EPSI21=-0.5/(ETAP(2)*ETAP(2))
         EPSI22=0.5*DETA2/(ETAP(1)*ETAP(2)*ETAP(2))
         H(1)=H(1)+EPSI16*G6*G6
         H(2)=H(2)-EPSI22*G2*G2-EPSI21*G1*G1
      ELSE IF (G1.GT.0. .AND. G2.LE.0.) THEN
         EPSI16=0.5/(ETAP(2)*ETAP(3))
         EPSI22=0.5*DETA2/(ETAP(1)*ETAP(2)*ETAP(2))
         H(1)=H(1)+EPSI16*G6*G6
         H(2)=H(2)-EPSI22*G2*G2
      ELSE IF (G2.GT.0. .AND. G3.LE.0.) THEN
         EPSI14=0.5*DETA1/(ETAP(1)*ETAP(1)*ETAP(3))
         EPSI13=-0.5/(ETAP(1)*ETAP(1))
         EPSI32=0.5/(ETAP(1)*ETAP(2))
         H(1)=H(1)-EPSI14*G4*G4-EPSI13*G3*G3
         H(3)=H(3)+EPSI32*G2*G2
      ELSE IF (G3.GT.0. .AND. G4.LE.0.) THEN
         EPSI14=0.5*DETA1/(ETAP(1)*ETAP(1)*ETAP(3))
```

(cont.)

```
      EPSI32=0.5/(ETAP(1)*ETAP(2))
      H(1)=H(1)-EPSI14*G4*G4
      H(3)=H(3)+EPSI32*G2*G2
   ELSE IF (G4.GT.0. .AND. G5.LE.0.) THEN
      EPSI24=0.5/(ETAP(1)*ETAP(3))
      EPSI35=-0.5/(ETAP(3)*ETAP(3))
      EPSI36=0.5*DETA3/(ETAP(2)*ETAP(3)*ETAP(3))
      H(2)=H(2)+EPSI24*G4*G4
      H(3)=H(3)-EPSI36*G6*G6-EPSI35*G5*G5
   ELSE
      EPSI24=0.5/(ETAP(1)*ETAP(3))
      EPSI36=0.5*DETA3/(ETAP(2)*ETAP(3)*ETAP(3))
      H(2)=H(2)+EPSI24*G4*G4
      H(3)=H(3)-EPSI36*G6*G6
   END IF
   PHIH(1)=SKP(ZX(2),ZY(2),U(1),V(1))
   PHIH(2)=SKP(ZX(3),ZY(3),U(2),V(2))
   PHIH(3)=SKP(ZX(1),ZY(1),U(3),V(3))
   PHIK(1)=SKP(ZX(3),ZY(3),U(1),V(1))
   PHIK(2)=SKP(ZX(1),ZY(1),U(2),V(2))
   PHIK(3)=SKP(ZX(2),ZY(2),U(3),V(3))
   PHIM(1)=Z(3)-Z(2)
   PHIM(2)=Z(1)-Z(3)
   PHIM(3)=Z(2)-Z(1)
   DO 10 I=1,3
      DELTA(I)=PHIH(I)-PHIM(I)
10 CONTINUE
   H(1)=DELTA(3)*ETA(1)*ETA(2)+(PHIK(1)+DELTA(1)-
&        PHIM(1))*H(1)
   H(2)=DELTA(1)*ETA(2)*ETA(3)+(PHIK(2)+DELTA(2)-
&        PHIM(2))*H(2)
   H(3)=DELTA(2)*ETA(3)*ETA(1)+(PHIK(3)+DELTA(3)-
&        PHIM(3))*H(3)
   DO 20 I=1,3
      PZ=PZ+Z(I)*ETA(I)+H(I)
20 CONTINUE
   RETURN
ELSE
   IER=-1
   RETURN
END IF
END
```

Figure 10.5. Program listing of TVAL6.

```
SUBROUTINE TVAL12(PX,PY,X,Y,Z,ZX,ZY,PZ,IER)
INTEGER IER
REAL PX,PY,X(3),Y(3),Z(3),ZX(3),ZY(3),PZ
REAL SKP
REAL U(3),V(3),XP(3),YP(3),FA(3),FN(3),SL(3),
&     ZP(3),ZXP(3),ZYP(3),ETA(3),G(3),H(3),
&     PHIH(3),PHIK(3),PHIM(3),DELTA(3)
SKP(XX,YY,UU,VV)=XX*UU+YY*VV
IER=0
PZ=0.0
```

(cont.)

```
      XP(1)=0.5*(X(2)+X(3))
      XP(2)=0.5*(X(3)+X(1))
      XP(3)=0.5*(X(1)+X(2))
      YP(1)=0.5*(Y(2)+Y(3))
      YP(2)=0.5*(Y(3)+Y(1))
      YP(3)=0.5*(Y(1)+Y(2))
      IFLAG=1
      CALL HERMI2(0.5,X(2),Y(2),X(3),Y(3),Z(2),ZX(2),ZY(2),
     &          Z(3),ZX(3),ZY(3),IFLAG,ZP(1),FA(1),IERR)
      CALL HERMI2(0.5,X(3),Y(3),X(1),Y(1),Z(3),ZX(3),ZY(3),
     &          Z(1),ZX(1),ZY(1),IFLAG,ZP(2),FA(2),IERR)
      CALL HERMI2(0.5,X(1),Y(1),X(2),Y(2),Z(1),ZX(1),ZY(1),
     &          Z(2),ZX(2),ZY(2),IFLAG,ZP(3),FA(3),IERR)
      U(1)=X(3)-X(2)
      U(2)=X(1)-X(3)
      U(3)=X(2)-X(1)
      V(1)=Y(3)-Y(2)
      V(2)=Y(1)-Y(3)
      V(3)=Y(2)-Y(1)
      DN1=SKP(ZX(2),ZY(2),-V(1),U(1))
      DN2=SKP(ZX(3),ZY(3),-V(1),U(1))
      FN(1)=0.5*(DN1+DN2)
      DN1=SKP(ZX(3),ZY(3),-V(2),U(2))
      DN2=SKP(ZX(1),ZY(1),-V(2),U(2))
      FN(2)=0.5*(DN1+DN2)
      DN1=SKP(ZX(1),ZY(1),-V(3),U(3))
      DN2=SKP(ZX(2),ZY(2),-V(3),U(3))
      FN(3)=0.5*(DN1+DN2)
      DO 10 I=1,3
         SL(I)=SKP(U(I),V(I),U(I),V(I))
         ZXP(I)=SKP(FA(I),FN(I),U(I),-V(I))/SL(I)
         ZYP(I)=SKP(FA(I),FN(I),V(I),U(I))/SL(I)
 10   CONTINUE
      CALL COORDS(PX,PY,XP(1),XP(2),XP(3),YP(1),YP(2),YP(3),
     &          ETA,IERR)
      IF (IERR.EQ.1) THEN
         IER=-1
         RETURN
      END IF
      G(1)=ETA(3)-ETA(1)
      G(2)=ETA(2)-ETA(1)
      G(3)=ETA(2)-ETA(3)
      H(1)=-0.5*ETA(3)*ETA(3)
      H(2)=-0.5*ETA(1)*ETA(1)
      H(3)=-0.5*ETA(2)*ETA(2)
      IF (G(1).GT.0.) THEN
         IF (G(2).GT.0.) THEN
            IF (G(3).GT.0.) THEN
               H(3)=H(3)+0.5*G(2)*G(2)
            ELSE
               H(1)=H(1)+0.5*G(3)*G(3)
               H(3)=H(3)+0.5*G(2)*G(2)
            END IF
         ELSE
            H(1)=H(1)+0.5*G(3)*G(3)
         END IF
      ELSE IF (G(3).GT.0.) THEN
```

(cont.)

```
          IF (G(2).GT.0.) THEN
              H(2)=H(2)+0.5*G(1)*G(1)
              H(3)=H(3)+0.5*G(2)*G(2)
          ELSE
              H(2)=H(2)+0.5*G(1)*G(1)
          END IF
      ELSE
          H(1)=H(1)+0.5*G(3)*G(3)
          H(2)=H(2)+0.5*G(1)*G(1)
      END IF
      U(1)=XP(3)-XP(2)
      V(1)=YP(3)-YP(2)
      U(2)=XP(1)-XP(3)
      V(2)=YP(1)-YP(3)
      U(3)=XP(2)-XP(1)
      V(3)=YP(2)-YP(1)
      PHIH(1)=SKP(ZXP(2),ZYP(2),U(1),V(1))
      PHIH(2)=SKP(ZXP(3),ZYP(3),U(2),V(2))
      PHIH(3)=SKP(ZXP(1),ZYP(1),U(3),V(3))
      PHIK(1)=SKP(ZXP(3),ZYP(3),U(1),V(1))
      PHIK(2)=SKP(ZXP(1),ZYP(1),U(2),V(2))
      PHIK(3)=SKP(ZXP(2),ZYP(2),U(3),V(3))
      PHIM(1)=ZP(3)-ZP(2)
      PHIM(2)=ZP(1)-ZP(3)
      PHIM(3)=ZP(2)-ZP(1)
      DO 20 I=1,3
          DELTA(I)=PHIH(I)-PHIM(I)
20    CONTINUE
      H(1)=DELTA(3)*ETA(1)*ETA(2)+(PHIK(1)+DELTA(1)-
     &     PHIM(1))*H(1)
      H(2)=DELTA(1)*ETA(2)*ETA(3)+(PHIK(2)+DELTA(2)-
     &     PHIM(2))*H(2)
      H(3)=DELTA(2)*ETA(3)*ETA(1)+(PHIK(3)+DELTA(3)-
     &     PHIM(3))*H(3)
      DO 30 I=1,3
          PZ=PZ+ZP(I)*ETA(I)+H(I)
30    CONTINUE
      IF (ETA(1).LT.0.0) THEN
          PZ1=ZP(2)+ZP(3)-ZP(1)
     &        -DELTA(3)-0.5*(PHIK(1)+DELTA(1)-PHIM(1))
     &        +DELTA(1)-0.5*(PHIK(2)+DELTA(2)-PHIM(2))
     &        -DELTA(2)+1.5*(PHIK(3)+DELTA(3)-PHIM(3))
          PZ=PZ+(Z(1)-PZ1)*ETA(1)*ETA(1)
      ELSE IF (ETA(2).LT.0.0) THEN
          PZ2=ZP(1)+ZP(3)-ZP(2)
     &        -DELTA(3)+1.5*(PHIK(1)+DELTA(1)-PHIM(1))
     &        -DELTA(1)-0.5*(PHIK(2)+DELTA(2)-PHIM(2))
     &        +DELTA(2)-0.5*(PHIK(3)+DELTA(3)-PHIM(3))
          PZ=PZ+(Z(2)-PZ2)*ETA(2)*ETA(2)
      ELSE IF (ETA(3).LT.0.0) THEN
          PZ3=ZP(1)+ZP(2)-ZP(3)
     &        +DELTA(3)-0.5*(PHIK(1)+DELTA(1)-PHIM(1))
     &        -DELTA(1)+1.5*(PHIK(2)+DELTA(2)-PHIM(2))
     &        -DELTA(2)-0.5*(PHIK(3)+DELTA(3)-PHIM(3))
          PZ=PZ+(Z(3)-PZ3)*ETA(3)*ETA(3)
      END IF
      RETURN
      END
```

Figure 10.6. Program listing of TVAL12.

```
      SUBROUTINE EXPOL(N,PX,PY,I1,X,Y,X,ZXZY,IADJ,IEND,PZ,IER)
      INTEGER N,I1,IEND(N),IADJ(1),IER
      REAL PX,PY,X(N),Y(N),Z(N),ZXZY(2,N),PZ
      PZ=0.0
      N2=I1
10    N1=IADJ(IEND(N2)-1)
      X1=X(N1)
      Y1=Y(N1)
      X2=X(N2)
      Y2=Y(N2)
      U=X2-X1
      V=Y2-Y1
      DP1=-U*(PX-X2)-V*(PY-Y2)
      DP2=(PX-X1)*U+(PY-Y1)*V
      IF (DP1.LE.0.) THEN
          PZ=Z(N2)+ZXZY(1,N2)*(PX-X2)+ZXZY(2,N2)*(PY-Y2)
          IER=1
          RETURN
      ELSE IF (DP2.GT.0.) THEN
          ETA1=DP1/(U*U+V*V)
          ETA2=1.0-ETA1
          QX=ETA1*X1+ETA2*X2
          QY=ETA1*Y1+ETA2*Y2
          IFLAG=0
          CALL HERMI2(ETA2,X1,Y1,X2,Y2,Z(N1),ZXZY(1,N1),
     &                ZXZY(2,N1),Z(N2),ZXZY(1,N2),ZXZY(2,N2),
     &                IFLAG,PZ,PA,IER)
          PZ=PZ+(ETA1*ZXZY(1,N1)+ETA2*ZXZY(1,N2))*(PX-QX)+
     &          (ETA1*ZXZY(2,N1)+ETA2*ZXZY(2,N2))*(PY-QY)
          IER=1
          RETURN
      ELSE
          N2=N1
          GOTO 10
      END IF
      END
```

Figure 10.7. Program listing of EXPOL.

```
     SUBROUTINE HERMI2(ETA,X1,Y1,X2,Y2,F1,FX1,FY1,F2,FX2,FY2,
&                     IFLAG,F,FA,IER)
     INTEGER IFLAG,IER
     REAL ETA,X1,Y1,X2,Y2,F1,FX1,FY1,F2,FX2,FY2,F,FA
     REAL SKP
     SKP(X,Y,U,V)=X*U+Y*V
     IER=0
     F=0.0
     FA=0.0
     IF (IFLAG.EQ.0 .OR. IFLAG.EQ.1) THEN
         U=X2-X1
         V=Y2-Y1
         D=F2-F1
         ETAM1=ETA-1.0
         A1=SKP(FX1,FY1,U,V)
         B1=SKP(FX2,FY2,U,V)
         A2=-0.5*B1-1.5*A1+D+D
         B2= 1.5*B1+0.5*A1-D-D
         IF (ETA.LT.0.5) THEN
             F=ETA*(ETA*A2+A1)+F1
         ELSE
             F=ETAM1*(ETAM1*B2+B1)+F2
         END IF
         IF (IFLAG.EQ.1) THEN
             IF (ETA.LE.0.5) THEN
                 FA=2.0*A2*ETA+A1
             ELSE
                 FA=2.0*B2*ETAM1+B1
             END IF
         END IF
         RETURN
     ELSE
         IER=-1
         RETURN
     END IF
     END
```

Figure 10.8. Program listing of HERMI2.

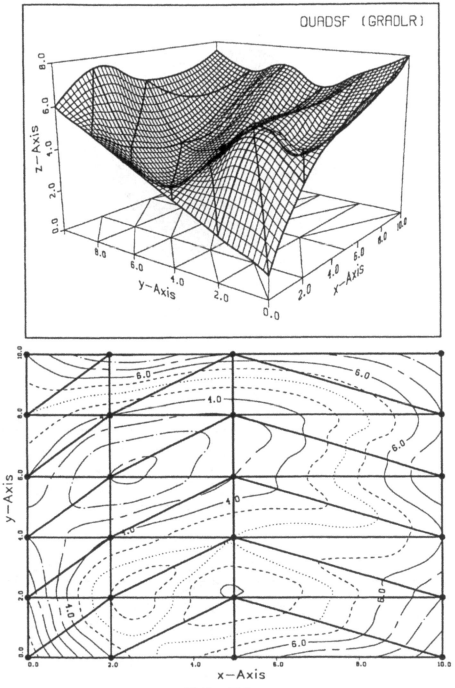

Figure 10.9.

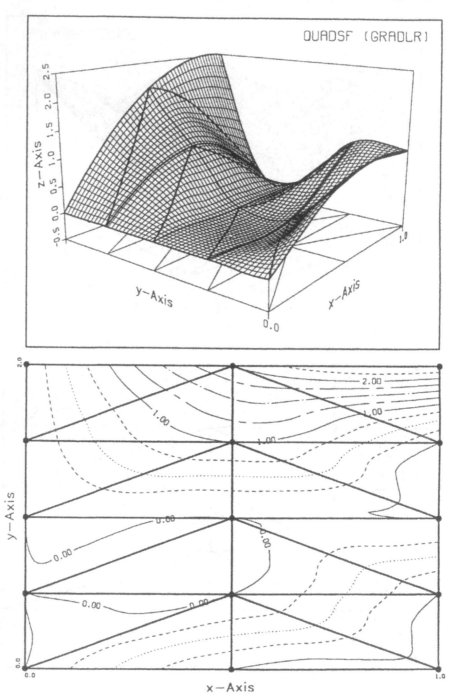

Figure 10.10.

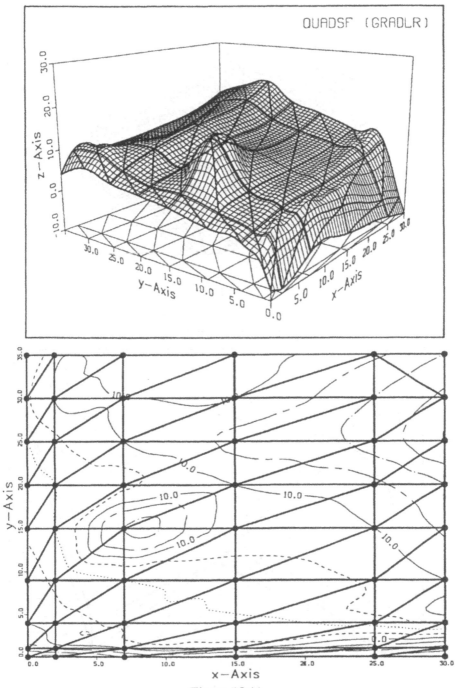

Figure 10.11.

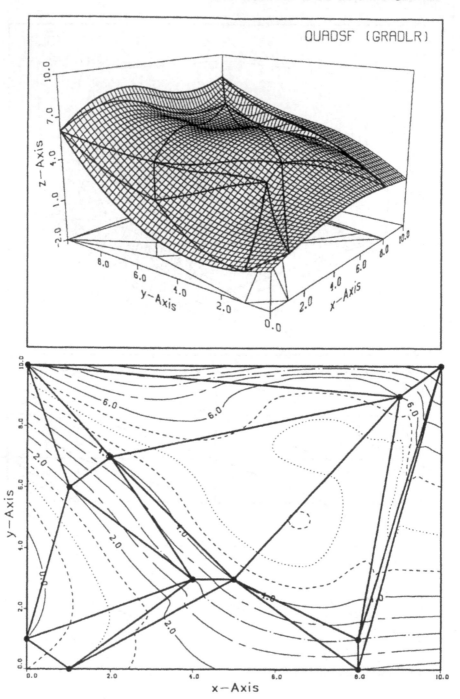

Figure 10.12.

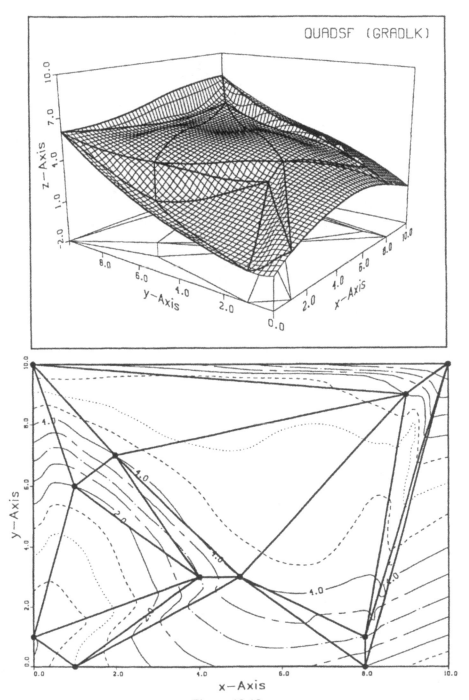

Figure 10.13.

11

Cubic Spline Interpolation over Triangulations

In order to obtain C^1 cubic spline interpolants over triangulations, it suffices to split each triangle into only three subtriangles. This smaller number is certainly an improvement but we will not be able to make use of an extension trick such as (10.2). On the other hand, the formulas are somewhat simpler ([59,68,81]).

We first consider just a single triangle $P_1P_2P_3$ of the triangulation. An arbitrary point S (which we will later take to be the centroid) inside the triangle splits it, according to Fig. 11.1 into three subtriangles, P_1SP_2, P_2SP_3, and P_3SP_1. On these we set three cubic polynomials q_3, q_1, and q_2 of the form (6.3) with $m = 3$. The 30 coefficients in total (actually, just as in Chapter 10, we will not give *explicit* formulas for these) are to be so determined that values z_i and values zx_i, zy_i for the first partial derivatives, either given or estimated, are interpolated at the vertices P_1, P_2, and P_3, and so that the function q on $P_1P_2P_3$, formed from q_1, q_2, and q_3 is once continuously differentiable. Further, we ask that the global function Q, formed from such $q's$ on each of the triangles of the triangulation, is in its turn continuously differentiable on the whole triangulation. As in Chapter 10, the construction on each individual triangle is such that this happens automatically.

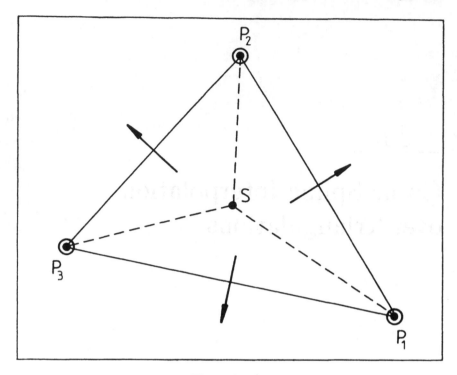

Figure 11.1.

Of central importance is the fact that a cubic polynomial and its gradient
are uniquely determined on an edge of an arbitrary triangle by its function
values and first partial derivtives at the vertices as well as the value of
the normal derivative at an interior point of the edge ([68]). If one now
arranges that on each edge the normal derivative is a linear function of
the two normal derivatives at the vertices, then continuous differentiability
follows, both for the subtriangles and for the triangles of the triangulation.
The easiest thing to do is, for example, on the side P_1P_2, to assign the
value,

$$\left. \frac{\partial q}{\partial \mathbf{n}} \right|_{H_3} = \frac{1}{2} \left[\left. \begin{pmatrix} zx \\ zy \end{pmatrix} \right|_{P_1} + \left. \begin{pmatrix} zx \\ zy \end{pmatrix} \right|_{P_2} \right] \cdot \mathbf{n}, \qquad (11.1)$$

at the edge midpoint H_3 ($\mathbf{n}$ is a vector perpendicular to the side). This
then requires only estimates of the gradients (and not of higher derivatives
as in the next chapter). In Fig. 11.1, the normal derivatives are indicated
by arrows. The single circles at the vertices are to mean that first partial
derivatives must be available there.

Without wanting to go into much detail, we will at least verify that there are sufficiently many conditions to match the 30 coefficients per triangle ([73]). There are six interpolation conditions each on the exterior edges for the three cubics defined on the subtriangles of a triangle $P_1P_2P_3$, which gives a total of 18. In addition to these, there are the three conditions on the normal derivatives on the exterior edges. Further, the three cubics must have the same function values and gradients at S (see Fig. 11.1), which gives six conditions. Finally, the normal derivatives must agree at a point on each of the three inner seams of $P_1P_2P_3$ emanating from S, which gives the remaining three of the necessary 30 conditions. These conditions are also sufficient, i.e., the corresponding linear system of equations is non-singular ([68,73]).

As already mentioned, the choice of S as the centroid results in the three subtriangles having equal areas and is in this sense equally weighted. Moreover, if the normal derivative values are also set according to (11.1) then the setting up and solution of a linear system is not necessary: it breaks down into easily solved equations. There are also relatively simple formulas (in barycentric coordinates) for the evaluation of q ([59,60,68,81]).

The complete method is implemented in the subroutine INTRC1, which is part of Renka's software package ([84,85]). There is also an arrangement for extrapolation oustide the convex hull of the given points ([68,81,85]). INTRC1 makes standard use of GRADL, which also belongs to the package, to estimate gradients. Alternatively, GRADG is also recommended. We have also carried out the example calculations with the subroutines GRADLA, GRADLK, and GRADH given in Chapter 9. In choosing the examples, we have tried to take typical effects into consideration while being aware that no objective measures could have been applied to decide this.

Figures 11.2–11.4 (cf. Figs. 1.1, 3.6, 3.21, 4.9, 4.22, 5.18, 6.10–6.12, 8.7) show the same example with the different gradient estimators, GRADL, GRADG, and GRADH. Figure 11.5 shows an example hitherto not used. In Fig. 11.6 (cf. 1.3, 2.8, 3.23, 4.11, 5.19, 10.11), the surface is practically independent of the gradient estimator. GRADH was again used in Fig. 11.7 (cf. 1.4, 2.10, 3.19, 3.24, 4.14, 5.21). Comparing the effect of GRADL, GRADG, and GRADLK in Figs. 11.8–11.10 (cf. 2.13, 3.11, 3.25, 4.17, 4.25, 5.24, 8.4), GRADLK comes out better than GRADG, and GRADG better than GRADL. These should especially be compared with Fig. 4.25. Next, we give some examples that were not used in Part I and for which the extrapolation effect is also visible in part. Figures 11.11 and 11.12 show the different results using GRADG and GRADH on an example that has not yet been used. Figures 11.13 (cf. 6.21, 8.6) and 11.14 (cf. 6.22, 8.7) have slightly different node sets. Figures 11.15 (cf. 6.23) and 11.16 (cf. 6.25, 8.9) have the same node set but partly different z values. Finally,

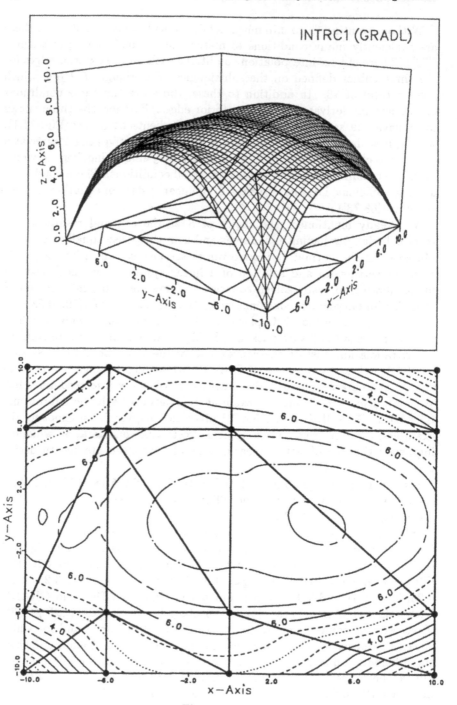

Figure 11.2.

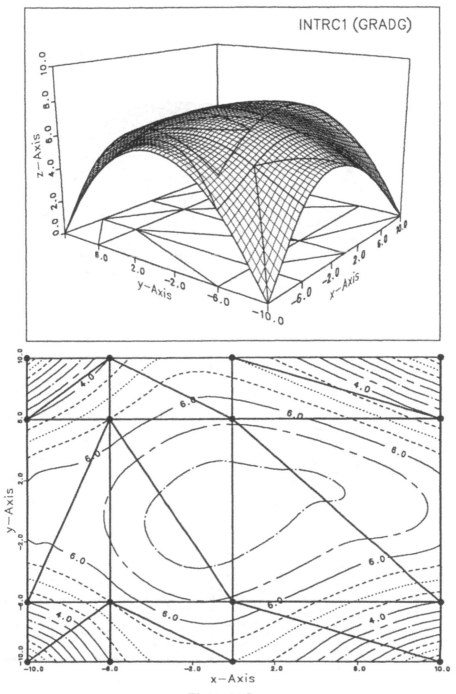

Figure 11.3.

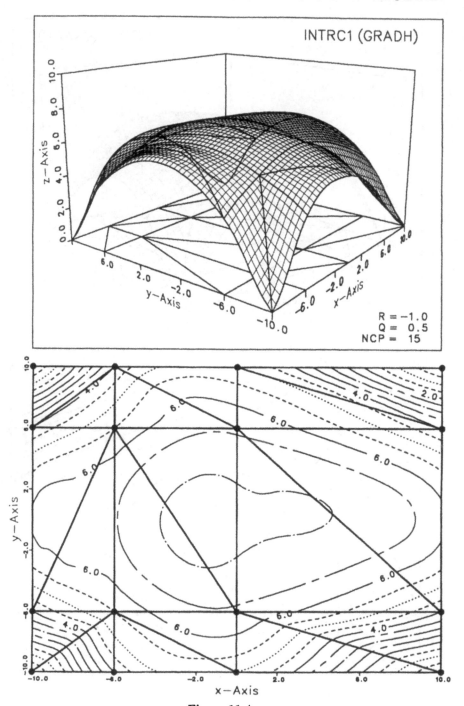

Figure 11.4.

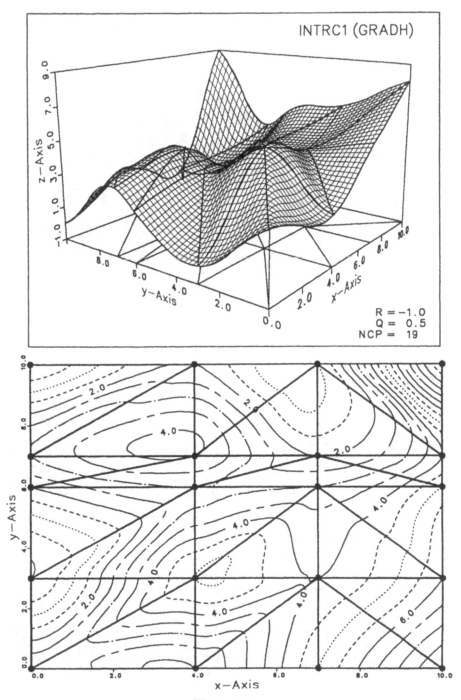

Figure 11.5.

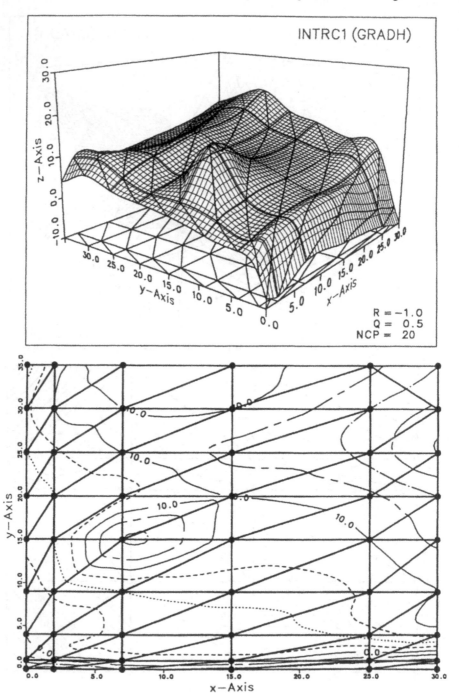

Figure 11.6.

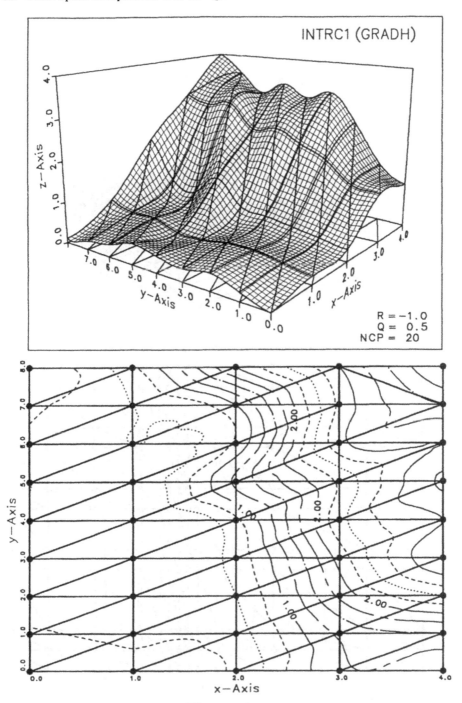

Figure 11.7.

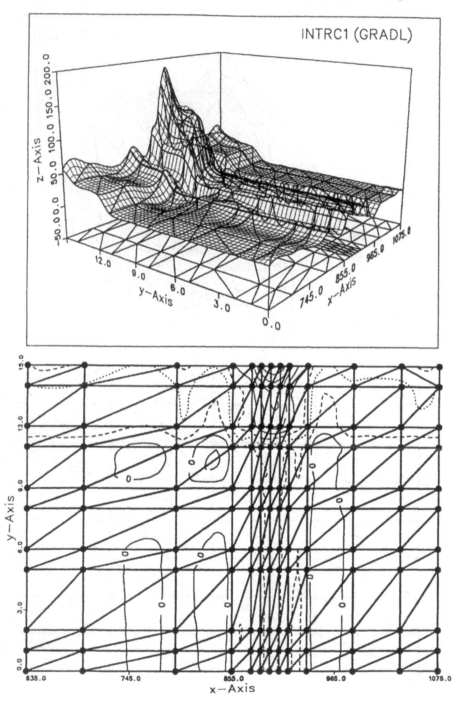

Figure 11.8.

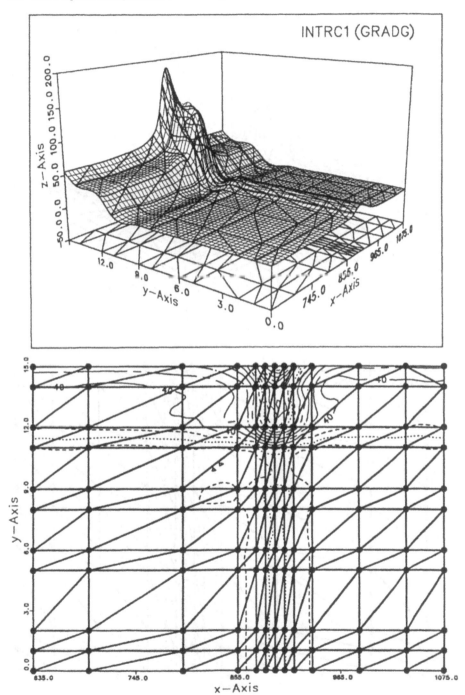

Figure 11.9.

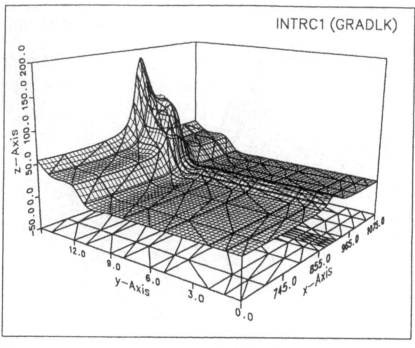

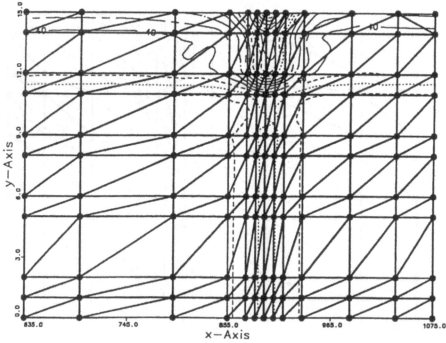

Figure 11.10.

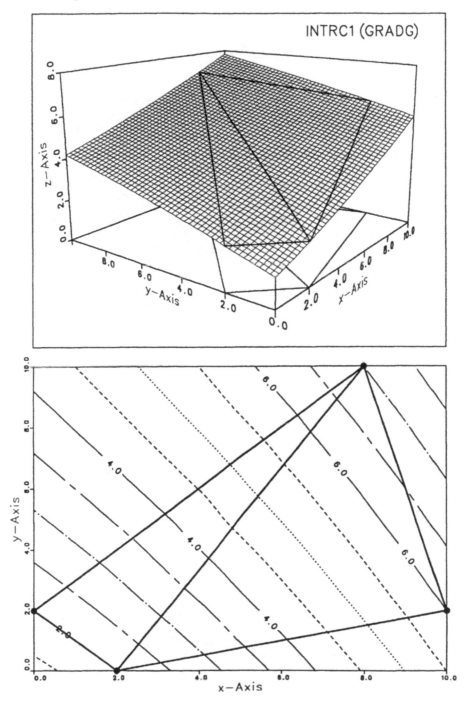

Figure 11.11.

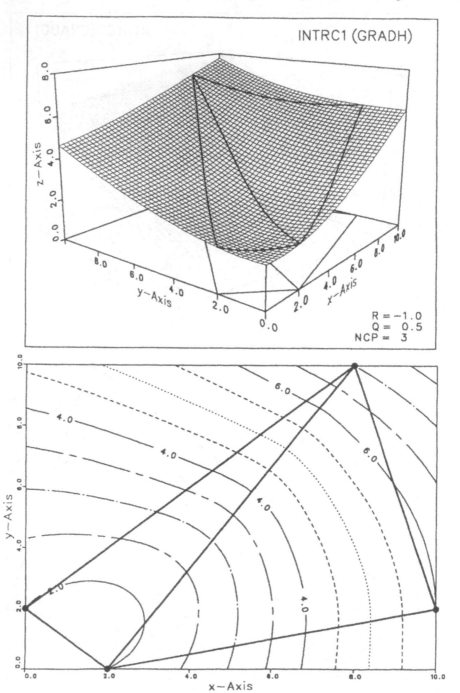

Figure 11.12.

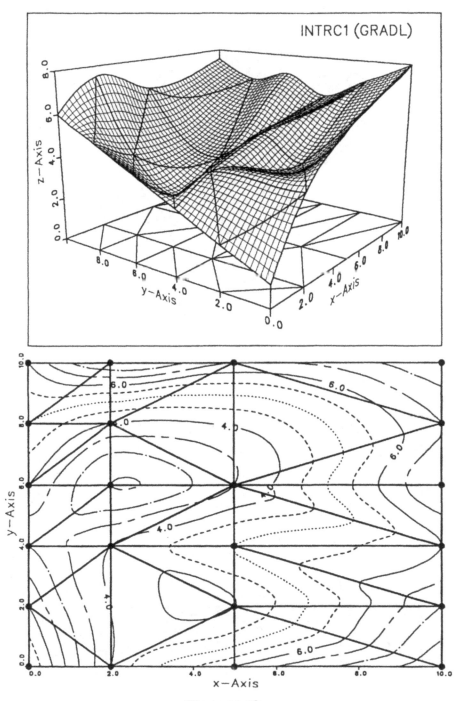

Figure 11.13.

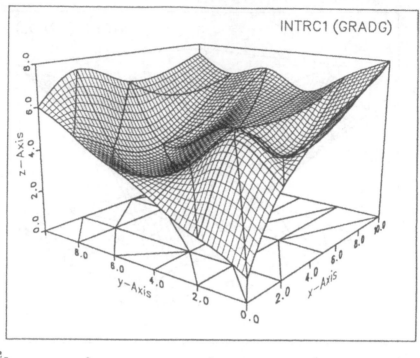

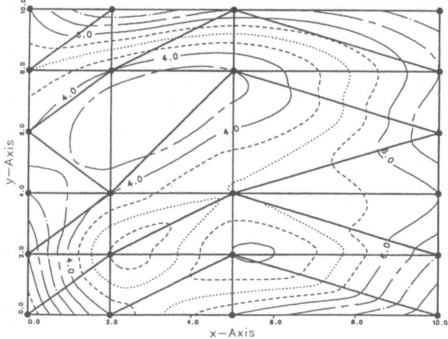

Figure 11.14.

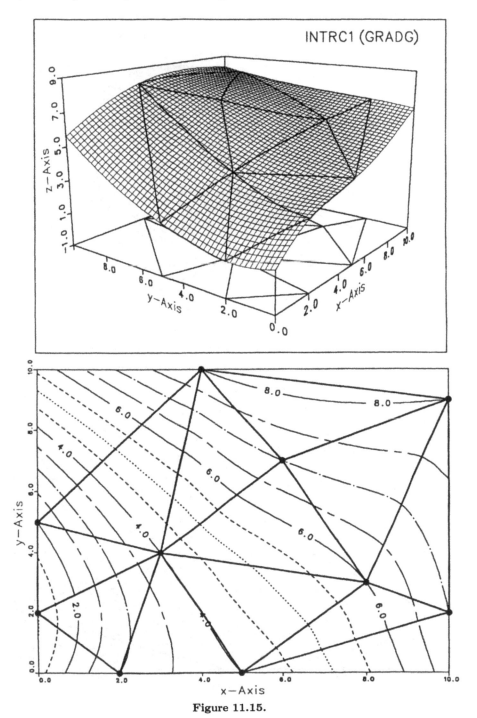

Figure 11.15.

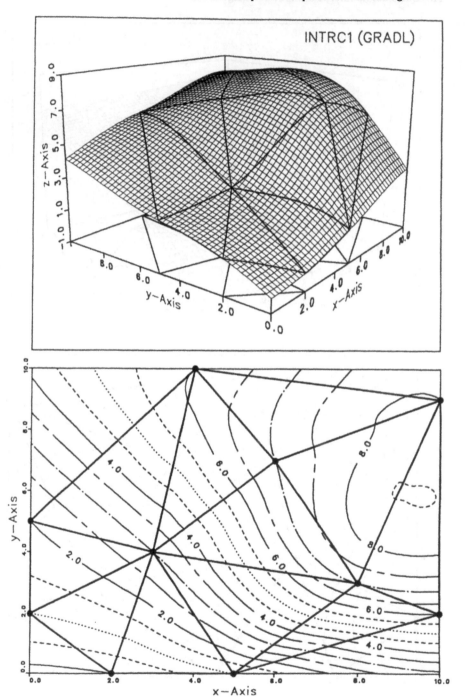

Figure 11.16.

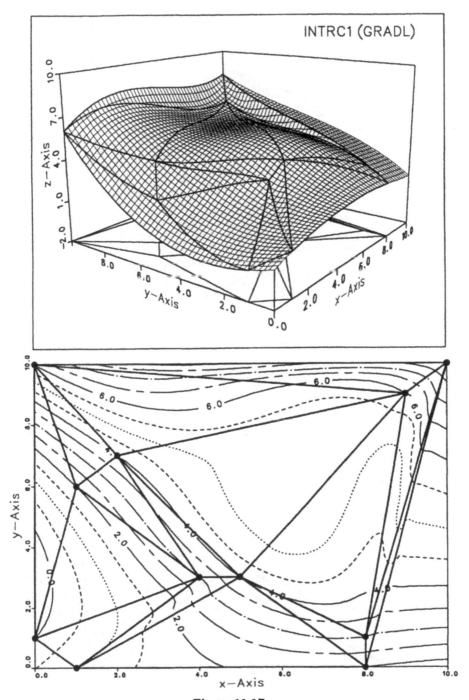

Figure 11.17.

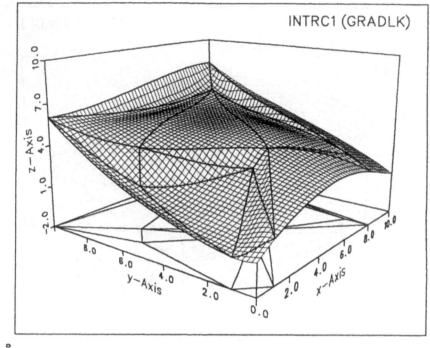

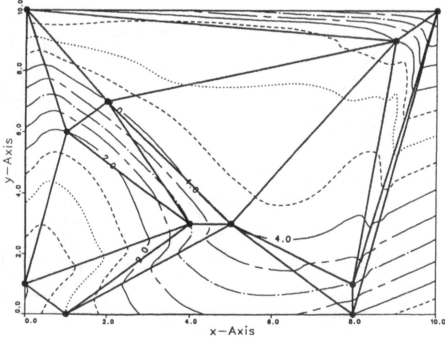

Figure 11.18.

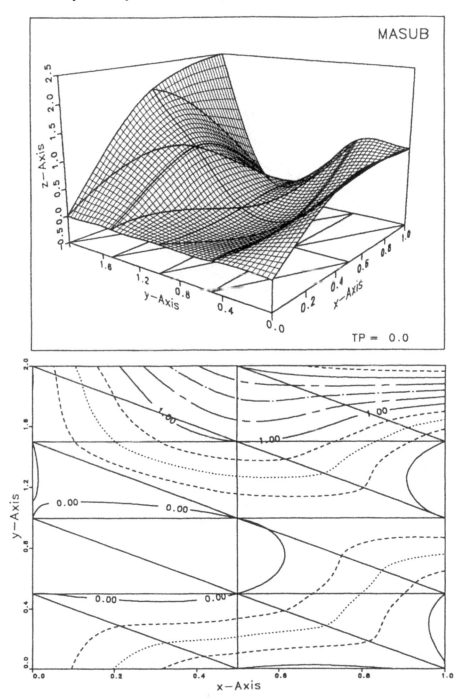

Figure 11.19.

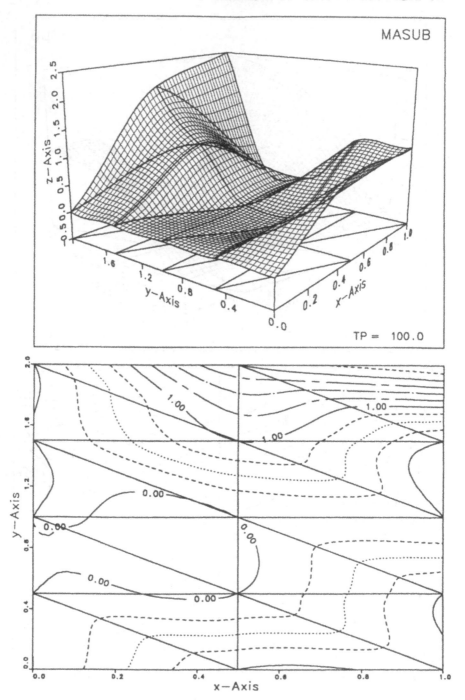

Figure 11.20.

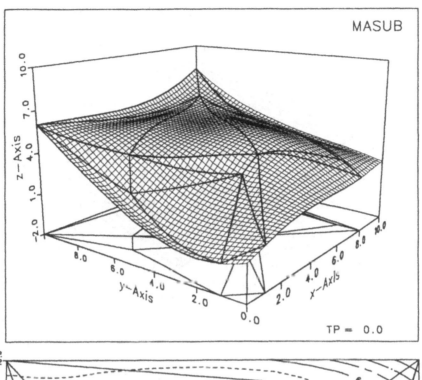

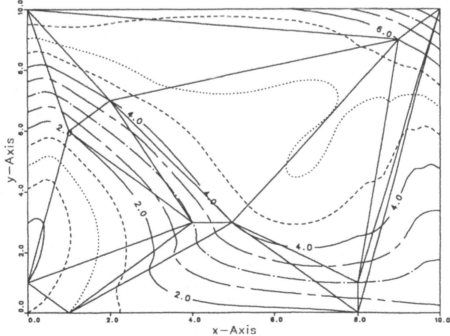

Figure 11.21.

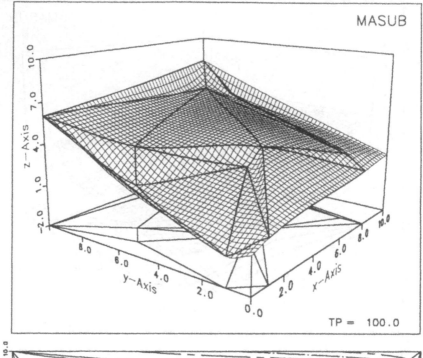

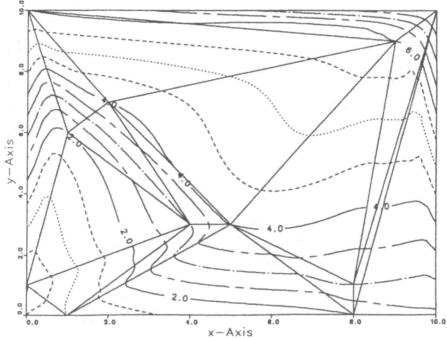

Figure 11.22.

Figs. 11.17 and 11.18 (cf. 6.26, 8.10, 10.12, 10.13), where the Delaunay triangulation has some very obtuse triangles, again show the influence of two different gradient estimators.

A software package similar to that of Renka (subroutine MASUB and auxiliary routines, main test program) is described in [67]. The triangulation is computed with the subroutine of Akima (from Algorithm 576) using Lawson's method. The C^1 spline interpolant is calculated using a discretized version of Nielson's 9-parameter interpolant ([69]), which is based on blending and does not require a subdivision of the triangle. The necessary gradient estimates are calculated by means of the iterative minimization of a convenient functional that can be adjusted by the choice of a tension parameter $TP \geq 0$. If so desired, the surface can be extrapolated to a surrounding rectangle using Shepard's method. The advantage of the method is probably that undesired oscillations can be suppressed through the choice of a large value for TP (global for all nodes).

Figures 11.19 and 11.20 (cf. Figs. 1.2, 2.7, 3.7, 3.22, 4.10, 4.23, 6.13–6.15, 8.2, 10.10) show the results of MASUB for the same example, first with $TP = 0$ and then $TP = 100$. The same holds for Figs. 11.21 and 11.22 (cf. 6.26, 8.10, 10.12, 10.13, 11.17, 11.18) on another example for which extrapolation was required.

In these and other examples that we computed, no significant differences between MASUB and HARDY, QUADSF or INTRC1 were detectable. Hence we see no reason to prefer MASUB over QUADSF or INTRC1. Execution times were not taken into consideration in making this statement. We prefer INTRC1 except when *contour lines* are to be calculated, when it may be more advantageous to use QUADSF. This also holds if we take the method of the next chapter into consideration.

It should be mentioned that for the triangle subdivision of Fig. 11.1, it is also possible to construct a C^1 spline interpolant of degree $m = 4$ ([73]).

To end, we mention the latest attempts ([79,80]) to use data-dependent triangulations not only for C^0 (see Chapter 7) but also for C^1 spline interpolants. Just as before, the results so far seem convincing only for special datasets.

12

C^1 Spline Interpolation of Degree Five on Triangulations

If one does not want to subdivide triangles into subtriangles, then one must increase both the polynomial degree *and* the number of interpolation conditions. Akima's method ([4]) assumes estimates for the first (zx_i, zy_i) *and* second partial derivatives $(zxx_i, zxy_i = zyx_i, zyy_i)$ at all the given points P_i, $i = 1, \cdots, n$. For each triangle, a *polynomial of degree five*, as in (6.3), that takes on these values as well as the z_i is then to be determined. This gives 18 conditions as opposed to 21 polynomial coefficients. The remaining three conditions come from requiring that the normal derivative restricts on each triangle edge to an only *cubic* polynomial in one variable in the direction of the edge in question. The situation is pictured in Fig. 12.1. The relevant details are proven in [4] or [3]. The asssociated global function is C^1 on the whole triangulation.

The first derivatives can be determined by the methods described in Chapter 9. The same methods may be applied to the already computed values zx_i and zy_i, to yield zxx_i, zxy_i and zyx_i, zyy_i, respectively. The values zxy_i and zyx_i are then averaged. For any particular point (x_i, y_i, z_i), only the NCP nearest neighbors (x_k, y_k, z_k), $k = 1, \cdots, \text{NCP}$, are used. $3 \le \text{NCP} \le 5$ is recommended ([4]).

The corresponding subroutine IDBVIP ([5]), available from IMSL or over NANET, extrapolates with a bivariate polynomial of degree five in the

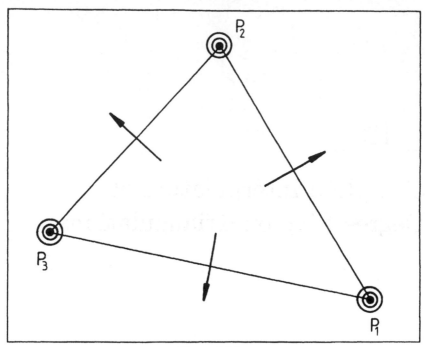

Figure 12.1.

variable in the direction of an exterior edge of the triangulation and of degree two in the second variable in its normal direction. Figures 12.2 (cf. Figs. 1.2, 4.10, 4.23, 6.13, 10.10, 11.19), 12.3 (cf. 6.25, 11.16), and 12.4 (cf. 6.26, 10.13, 11.17, 11.18, 11.21, 11.22) were produced in this way. When one compares these and other results with those of, say, INTRC1, there are often no serious differences. Surprisingly, we find that despite the higher degree, there seldom arise any unwanted oscillations in the interpolating surface.

In [6], Akima gives some criticisms of his own method. The user must assign NCP. The determination of the NCP closest points in $\mathbb{R}^2$ can be time-consuming. Then some modifications are suggested and compared. Among these, the setting of NCP is done away with, NCP is chosen to be $n - 1$, and for smoothing purposes, the cross products are supplied with a weight when they are summed (see Chapter 9). One of these variants (with NCP no longer a parameter) is implemented in the subroutine SURF ([54]). Since the documentation refers only to the old work ([4]), it is not clear exactly which variant it actually is.

Comparing the examples in Figs. 12.5 and 12.6, computed with SURF, with 12.2 and 12.4, we see some slight differences that do not argue in SURF's favor. Also, for other examples, IDBVIP seemed to produce a better-looking surface. We did not take execution time into consideration in this comparison.

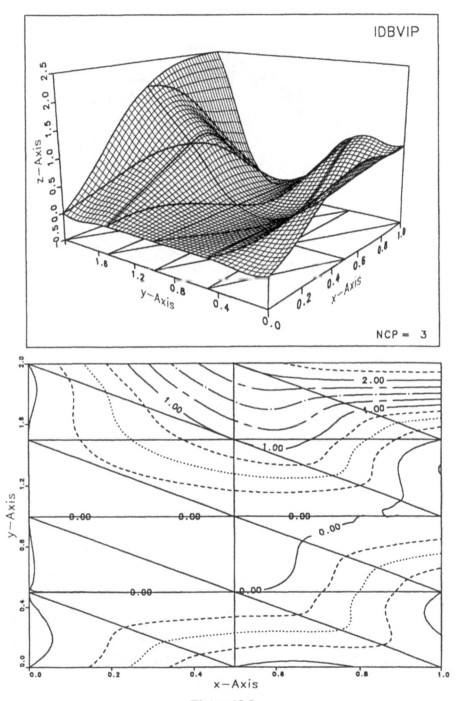

Figure 12.2.

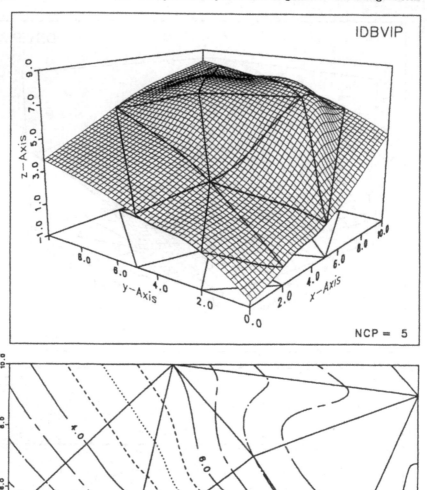

Figure 12.3.

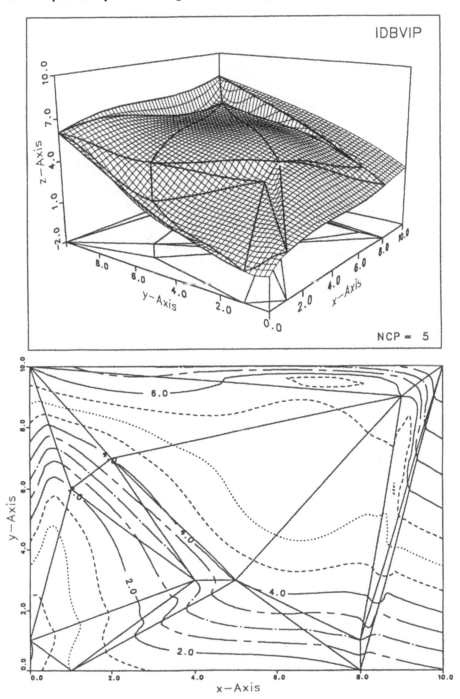

Figure 12.4.

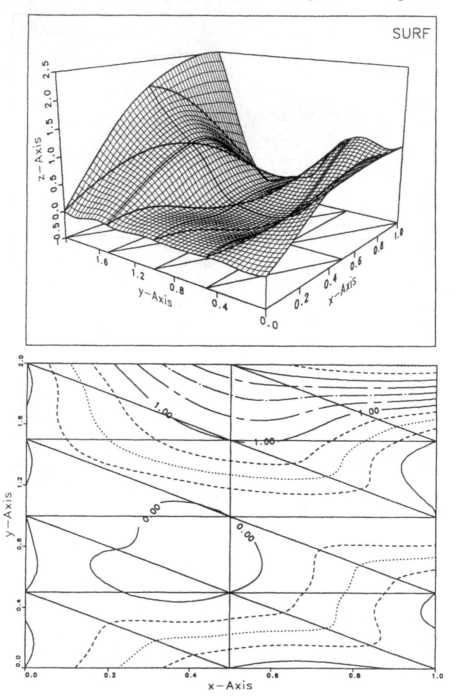

Figure 12.5.

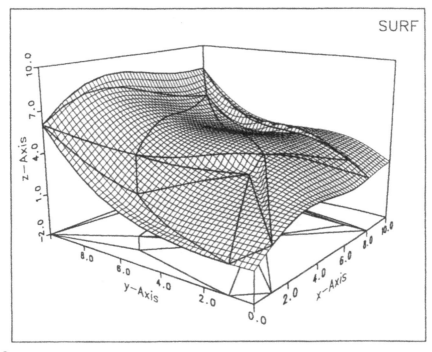

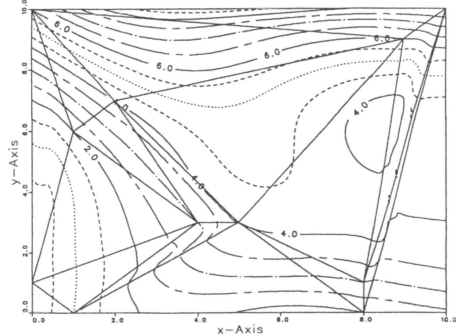

Figure 12.6.

Postscript

There was very little that could have been added to Part I. Hermite biquadratic and Hermite quadratic spline interpolation (subdivision of the rectangle into a total of 16 triangles) are the thesis topics of Gunnar Duis and Thomas Hillen at the University of Oldenburg. These were completed at the end of 1990 and give the relevant theoretical details, subroutines, and a number of examples. As regards bicubic spline interpolation on a polar grid, which seems to be important in optics, we refer the reader to [15,48,49,106]. There are also some newer works on shape preservation [19,33,105].

Part II, however, as the bibliography [42] and the most recent survey article [43] show, could have been substantially more extensive. In particular we mention the global C^1 method using cubic polynomials [92], which needs no derivative estimates. Likewise, thin plate splines ([38,40]) could also be important. Both methods have also been implemented. C^1 spline interpolation on spherical triangles on the surface of the sphere is treated in [61,82]; software is given in [83]. The spline interpolation problem for more than two variables is considered in [7]. We make special mention of the publications [15,24,31,32,35,36,37,39,41,46,71,72,74,94] that, for one reason or the other, appear to us to be especially interesting, but have no role to play in our choice of material. Shape preservation problems on triangulations do not seem to have been addressed yet.

A

Appendix

In Part I, the subroutine pair TRDISA (Figs. A.1 and A.2) and TRDISB (Fig. A.3) is used for solving a collection of linear systems having the same symmetric tridiagonal coefficient matrix but different right-hand sides. Similarly, the pair TRDIUA (Fig. A.4) and TRDIUB (Fig. A.5) is used if the coefficient matrix is nonsymmetric. TRDISA and TRDISB come from splitting the subroutine TRIDIS given in the appendix of [100] and TRDIUA and TRDIUB from splitting TRIDIU ([100]). The LU decomposition takes place in TRDISA or TRDIUA and then the solution for a particular right-hand side is done in TRDISB or TRDIUB. The reader may find the details of these methods in [100] and hence we give the subroutine listings and program descriptions without further comment.

```
       SUBROUTINE TRDISA(N,A,B,C,EPS,IFLAG)
       DIMENSION A(N),B(N),C(N)
       IFLAG=0
       N1=N-1
       DO 10 K=2,N
          KM1=K-1
          H=B(KM1)
          IF (ABS(H).LT.EPS) THEN
             IFLAG=2
             RETURN
          END IF
          H=A(KM1)/H
          C(KM1)=H
          B(K)=B(K)-H*A(KM1)
10     CONTINUE
       RETURN
       END
```

Figure A.1. Program listing of TRDISA.

Calling sequence:

CALL TRDISA(N,A,B,C,EPS,IFLAG)

Purpose:
Calculation of the LU decomposition of a symmetric tridiagonal matrix M, i.e., TRDISA computes matrices,

$$L = \begin{pmatrix} B(1) & & & \\ A(1) & B(2) & & \\ & \cdot & \cdot & \\ & & \cdot & \cdot \\ & & & A(N-1) & B(N) \end{pmatrix}$$

and

$$U = \begin{pmatrix} 1 & C(1) & & \\ & \cdot & \cdot & \\ & & \cdot & \cdot \\ & & 1 & C(N-1) \\ & & & 1 \end{pmatrix},$$

such that $M = LU$.

Description of the parameters:

N	Dimension of the matrix. Restriction: $N \geq 1$.
A	ARRAY(N): Input: Superdiagonal of M $(1, \cdots, N-1)$.
	Output: Subdiagonal of L.

B	ARRAY(N): Input: Diagonal of M.
	Output: Diagonal of L.
C	ARRAY(N): Output: Superdiagonal of U $(1, \cdots, N-1)$.
EPS	Value used for accuracy test. If the absolute value of a transformed element on the main diagonal is smaller than EPS, then execution is terminated with IFLAG=2. Recommendation: EPS$= 10^{-t+2}$, where t is the number of available decimal digits.
IFLAG	=0: Normal execution.
	=2: The matrix is numerically singular.

Figure A.2. Description of TRDISA.

```
      SUBROUTINE TRDISB(N,A,B,C,D)
      DIMENSION A(N),B(N),C(N),D(N)
      D(1)=D(1)/B(1)
      DO 10 K=2,N
         KM1=K-1
         D(K)=(D(K)-A(KM1)*D(KM1))/B(K)
10    CONTINUE
      DO 20 K=N-1,1,-1
         D(K)=D(K)-C(K)*D(K+1)
20    CONTINUE
      RETURN
      END
```

Calling sequence:

CALL TRDISB(N,A,B,C,D)

Purpose:
Calculation of the solution of a linear system of equations with symmetric tridiagonal coefficient matrix M. The LU decomposition of M must be precomputed by TRDISA. The vectors, A, B, and C, must correspond to the vectors, A, B, and C of TRDISA.

Description of the parameters:

N	Dimension of the matrix M. Restriction: $N \geq 1$.
A	ARRAY(N): Input: Vector A as computed by TRDISA.
B	ARRAY(N): Input: Vector B as computed by TRDISA.
C	ARRAY(N): Input: Vector C as computed by TRDISA.
D	ARRAY(N): Input: Right-hand side of the linear system.
	Output: Solution vector.

Figure A.3. Program listing of TRDISB and its description.

```
      SUBROUTINE TRDIUA(N,A,B,C,EPS,IFLAG)
      DIMENSION A(N),B(N),C(N)
      IFLAG=0
      IF (N.EQ.1) RETURN
      N1=N-1
      IF (ABS(B(1)).LT.EPS) THEN
         IFLAG=2
         RETURN
      END IF
      C(1)=C(1)/B(1)
      DO 10 K=2,N1
         KM1=K-1
         B(K)=B(K)-A(KM1)*C(KM1)
         IF (ABS(B(K)).LT.EPS) THEN
            IFLAG=2
            RETURN
         END IF
         C(K)=C(K)/B(K)
10    CONTINUE
      B(N)=B(N)-A(N1)*C(N1)
      RETURN
      END
```

Calling sequence:

CALL TRDIUA(N,A,B,C,EPS,IFLAG)

Purpose:
Calculation of the LU decomposition of a nonsymmetric tridiagonal matrix M.

Description of the parameters:

N,B,EPS,IFLAG as in TRDISA.
A   ARRAY(N): Input: Subdiagonal of M $(1, \cdots, N-1)$.
                      Output: Subdiagonal of L.
C   ARRAY(N): Input: Superdiagonal of M $(1, \cdots, N-1)$.
                      Output: Superdiagonal of U.

Figure A.4. Program listing of TRDIUA and description.

```
      SUBROUTINE TRDIUB(N,A,B,C,D)
      DIMENSION A(N),B(N),C(N),D(N)
      D(1)=D(1)/B(1)
      DO 10 K=2,N
         KM1=K-1
         D(K)=(D(K)-A(KM1)*D(KM1))/B(K)
10    CONTINUE
      DO 20 K=N-1,1,-1
         D(K)=D(K)-C(K)*D(K+1)
20    CONTINUE
      RETURN
      END
```

Calling sequence:

CALL TRDIUB(N,A,B,C,D)

Purpose:
Calculation of the solution of a linear system of equations with non-symmetric tridiagonal coefficient matrix M. The LU decomposition of M must be precomputed by TRDIUA. The vectors, A, B, and C, must correspond to the vectors A, B, and C of TRDIUA.

Description of the parameters:

N	Dimension of the matrix M. Restriction: $N \geq 1$.
A	ARRAY(N): Input: Vector A as computed by TRDIUA.
B	ARRAY(N): Input: Vector B as computed by TRDIUA.
C	ARRAY(N): Input: Vector C as computed by TRDIUA.
D	ARRAY(N): Input: Right-hand side of the linear system.
	Output: Solution vector.

Figure A.5. Program listing of TRDIUB and its description.

B

List of Subroutines

List of Subroutines				
No.	Name	Required Subroutines	Listing on Page	Description on Page
1	INTTWO		14	15
2	SBILIN	INTTWO	17	17
4	QUM2D	QUOPT2, QUMMAT	38	40
6	QUMVAL	INTTWO	42	42
7	QUA2D	QUADEC, QUPERM, QUAMAT, TRDISA, TRDISB	54	56
11	QBIVAL	INTTWO	59	59
12	CUB2D	CBPERM, CUBMAT TRDISA, TRDISB	75	77
15	CBIVAL	INTTWO	78	
16	HCUB2D	GRAD7, CUBMAT	96	98
17	QVASPL	CUB2D	111	112
18	QUAVAL	INTTWO	113	
19	RAT2D	RTPERM, RATMAT, TRDISA, TRDISB	127	129
22	RBIVAL	INTTWO, GI	131	131

No.	Name	Required Subroutines	Listing on Page	Description on Page
24	RSP2D	RSPDEC, RSPPER, RSPMAT, TRDIUA, TRDIUB	132	134
28	RSPVAL	INTTWO, GI	137	137
30	RVASPL	RSP2D	149	151
31	RVAVAL	INTTWO, GPIJ	151	152
33	SHEPG		168	169
34	HARDY	DECOMP, SOLVE	174	175
35	FHARDY		175	176
36	GRADH	HARDY, CLDP	216	217
38	GRADLK	TRLIST, INSERT	221	222
41	GRADLA	CLDP	225	226
42	QUADSF	INPOL, EXPOL, TVAL6 TRFIND, COORDS, TVAL12, HERMI2	234	235
48	TRDISA		286	286
49	TRDISB		287	287
50	TRDIUA		288	288
51	TRDIUB		289	289

Remark: The subroutines can be obtained on diskette from the author for a cover charge.

Bibliography

[1] Akima, H. "A method of bivariate interpolation and smooth surface fitting based on local procedures," *Comm. ACM* 17, 18–20 (1974).

[2] Akima, H. "Algorithm 474: Bivariate interpolation and smooth surface fitting based on local procedures," *Comm. ACM* 17, 26–31 (1974).

[3] Akima, H. "A method of bivariate interpolation and smooth surface fitting for values given at irregularly distributed points," *OT Rep. 75-70*, U.S. Govt. Printing Office, Washington, D.C., August 1975.

[4] Akima, H. "A method of bivariate interpolation and smooth surface fitting for irregularly distributed data points," *ACM TOMS* 4, 148–159 (1978).

[5] Akima, H. "Algorithm 526: Bivariate interpolation and smooth surface fitting for irregularly distributed data points," *ACM TOMS* 4, 160–164 (1978).

[6] Akima, H. "On estimating partial derivatives for bivariate interpolation for scattered data," *Rocky Mountain J. Math.* 14, 41–52 (1984).

[7] Alfeld, P. "Scattered data interpolation in three or more variables." In Lyche, T., and Schumaker, L. L. *Mathematical methods in Computer Aided Geometric Design*, 1–33, Academic Press, Boston, 1989.

[8] Barnhill, R. E., and Little, F. F. "Adaptive triangular quadrature," *Rocky Mountain J. Math.* 14, 53–75 (1984).

[9] Beatson, R. K., and Ziegler, Z. "Monotonicity preserving surface interpolation," *SIAM J. Numer. Anal.* 22, 401–411 (1985).

[10] Brunet, P. "Increasing the smoothness of bicubic spline surfaces," *CAGD* 2, 157–164 (1985).

[11] Carlson, R. E. "Shape preserving interpolation." In Mason, J. C., and Cox, M. S. (Eds.), *Algorithms for Approximation*, Clarendon Press, Oxford, 1987.

[12] Carlson, R. E., and Fritsch, F. N. "Monotone piecewise bicubic interpolation," *SIAM J. Numer. Anal.* 22, 386–400 (1985).

[13] Carlson, R. E., and Fritsch, F. N. "An algorithm for monotone piecewise bicubic interpolation," *SIAM J. Numer. Anal.* 26, 230–238 (1989).

[14] Chang, R. C., and Lee, R. C. T. "On the average length of Delaunay triangulations," *BIT* 24, 269–273 (1984).

[15] Chi, C. "Curvilinear bicubic spline fit interpolation scheme," *Optica Acta* 20, 979–993 (1973).

[16] Chui, C. K., and He, T.-X. "Bivariate C^1 quadratic finite elements and vertex splines," *Math. Comp.* 54, 169–187 (1990).

[17] Cline, A. K., and Renka, R. J. "A storage efficient method for construction of a Thiessen triangulation," *Rocky Mountain J. Math.* 14, 119–139 (1984).

[18] Costantini, P. "Algorithms for shape-preserving interpolation." In Schmidt, J. W., and Späth, H. (Eds.), *Splines in Numerical Analysis*, Akademie-Verlag, Berlin, 1989.

[19] Costantini, P., and Fontanella, F. "Shape-preserving bivariate interpolation," *SIAM J. Numer. Anal.* 27, 488-506 (1990).

[20] Costantini, P. "SPBSI1: A code for computing shape-preserving bivariate spline interpolation," *Report 216*, Dipartimento di Matematica, Università di Siena, (1989).

[21] de Boor, C. "Bicubic spline interpolation," *J. Math. Phys.* 41, 212–218 (1962).

[22] de Boor, C. *A Practical Guide to Splines*, Springer-Verlag, 1978.

[23] Dodd, S. L., McAllister, D. F., and Roulier, J. A. "Shape-preserving spline interpolation for specifying bivariate functions on grids," *IEEE Comp. Graphics Appl.* 3, 70–79 (1983).

[24] Dooley, J. C. "Two dimensional interpolation of irregularly spaced data using polynomial splines," *Phys. Earth. Planet. Inter.* 12, 180–187 (1976).

[25] Dyn, N., and Levin, L. "Iterative solution of systems originating from integral equations and surface interpolation," *SIAM J. Numer. Anal.* 20, 377–390 (1983).

[26] Dyn, N., Levin, L., and Rippa, S. "Numercial Procedures for surface fitting of scattered data by radial functions," *SIAM J. Sci. Stat. Comput.* 7, 639–659 (1986).

[27] Dyn, N., Levin, L., and Rippa, S. "Data dependent triangulations for piecewise linear interpolation," *IMA J. Numer. Anal.* 10, 137–154 (1990).

[28] Dyn, N. "Interpolation of scattered data by radial functions." In Chui, C. K., Schumaker, L. L., and Utreras, F. I. (Eds.), *Topics in Multivariate Approximation,* Academic Press, Orlando Fla., 1987.

[29] Farwig, R. "Rate of convergence of Shepard's global interpolation formula," *Math. Comp.* 46, 577–590 (1986).

[30] Floriani, L. de, and Dettori, G. "Surface interpolation methods over rectangular and triangular grids," *Méthodes numériques dans les sciences de l'ingénieur, 2e Congr., Int. G.A.M.N.I.,* 163–174, Paris, 1980.

[31] Foley, T. A. "Three-stage interpolation to scattered data," *Rocky Mountain J. Math.* 14, 141–149 (1984).

[32] Foley, T. A. "Scattered data interpolation and approximation with error bounds," *Comp. Aided Geom. Design* 3, 163–177 (1986).

[33] Fontanella, F. "Shape preserving surface interpolation." In Chui, C. K., Schumaker, L. L., and Utreras, F. I. (Eds.), *Topics in Multivariate Approximation,* Academic Press, Orlando Fla., 1987.

[34] Forsythe, G. E., Malcolm, M. A., and Moler C. B. *Computer Methods for Mathematical Computations,* Prentice Hall, 1977.

[35] Franke, R. "Locally determined smooth interpolation at irregularly spaced points in several variables," *J. Inst. Maths. Applics.* 19, 471–482 (1977).

[36] Franke, R., and Nielson, G. "Smooth interpolation of large data sets of scattered data," *Int. J. Numer. Methods Engin.* 15, 1691–1704 (1980).

[37] Franke, R. "Scattered data interpolation: Tests of some methods," *Math. Comp.* 38, 181–200 (1982).

[38] Franke, R. "Smooth interpolation of scattered data by local thin plate splines," *Comp. & Maths. with Appls.* 8, 273–281 (1982).

[39] Franke, R., and Nielson, G. "Surface approximation with imposed conditions." In Barnhill, R. E., and Boehm, W. (Eds.), *Surfaces in CAGD*, North Holland, 1983.

[40] Franke, R. "Thin plate splines with tension," *Comp. Aided Geom. Design* 2, 87–95 (1985).

[41] Franke, R. "Recent advances in the approximation of surfaces from scattered data." In Chui, C. K., Schumaker, L. L., and Utreras, F. I. (Eds.), *Topics in Multivariate Approximation*, Academic Press, Orlando Fla., 1987.

[42] Franke, R., and Schumaker, L. L. "A bibliography of multivariate approximation." In Chui, C. K., Schumaker, L. L., and Utreras, F. I. (Eds.), *Topics in Multivariate Approximation*, Academic Press, Orlando Fla., 1987.

[43] Franke, R., and Nielson, G. "Scattered data interpolation and applications: a tutorial and survey." In Hagen , H. and Roller, D. *Geometric Modelling: Methods and their Applications*, Springer-Verlag, 1990.

[44] Fritsch, F. N., and Carlson, R. E. "Monotonicity preserving bicubic interpolation: A progress report," *CAGD* 2, 117–121 (1985).

[45] Fritsch, F. N., and Carlson, R. E. "BIMOND3: Monotone Piecewise Bicubic Hermite Interpolation Code," UCID-21143, Computing and Mathematics Research Division, Lawrence Livermore National Laboratory, August 1987.

[46] Gmelig Meyling, R. H. J. "Approximation by cubic C^1–splines on arbitrary triangulations," *Numer. Math.* 51, 65–85 (1987).

[47] Gordon, W. J., and Wixom, J. A. "Shepard's method of 'metric interpolation' to bivariate and multivariate interpolation," *Math. Comp.* 32, 253–264 (1978).

[48] Grzanna, J. "Zweidimensionale Splineinterpolation über einem Polargitter," *J. Approx. Th.* 22, 189–201 (1978).

[49] Grzanna, J. "B-splines on polar meshes," Zentralinstitut für Optik und Spektroskopie, Preprint 90-2, Akademie der Wissenschaften der DDR, Preprint 90-2, Berlin, March 1990.

[50] Hänler, A., and Maess, G. "2D-Interpolation by biquadratic splines with minimal surface." In Schmidt, J. W., and Späth, H. (Eds.), *Splines in Numerical Analysis*, Akademie-Verlag, Berlin, 1989.

[51] Hardy, R. L. "Multiquadric equations of topogography and other irregular surfaces," *J. Geophys. Res.* 76, 1905–1915 (1971).

[52] Heindl, G. "Interpolation and approximation by piecewise quadratic C^1−functions of two variables." In Schempp, W., and Zeller, K. (Eds.), *Multivariate Approximation Theory*, Birkhäuser, Basel, 1979.

[53] Heindl, G. "Construction and applications of Hermite interpolating quadratic spline functions of two and three variables." In: Schempp, W., and Zeller, K. (Eds.), *Multivariate Approximation Theory III*, Birkhäuser, Basel, 1985.

[54] *IMSL User's Manual: FORTRAN Subroutines for Mathematical Applications*, Version 1.0, April 1987.

[55] Isaacson, E., and Keller, H. B. *Analyse numerischer Verfahren*, Verlag Harri Deutsch, Frankfurt, 1973.

[56] Kansa, E. I. "Applications of Hardy's multiquadric interpolation to hydrodynamics," *Proc. Simulation Conf., San Diego, Calif.* 4, 111–116 (1986).

[57] Klucewicz, I. M. "A piecewise C^1 interpolant to arbitrarily spaced data," *Comp. Graph. Image Proc.* 8, 92–112 (1978).

[58] Lancaster, P., and Salkauskas, K. *Curve and Surface Fitting*, Academic Press, London, 1986.

[59] Lawson, C. L. "C^1−compatible interpolation over a triangle," TM 33-770, Jet Propulsion Laboratory, 1976.

[60] Lawson, C. L. "Software for C^1 surface interpolation." In Rice, J. R. (Ed.), *Mathematical Software III*, 161–194, Academic Press 1977.

[61] Lawson, C. L. "C^1 surface interpolation for scattered data on a sphere," *Rocky Mountain J. Math.* 14, 177–202 (1984).

[62] Little, F. F. "Convex combination surfaces." In Barnhill, R. E., and Boehm, W. (Eds.), *Surfaces in CAGD*, 99–107, North Holland 1983.

[63] Maess, G. "Smooth interpolation of curves and surfaces by quadratic splines with minimal curvature." In Sendov, B., Lazarov, R., and Vasilevski, P. (Eds.), *Proc. Internat. Conf. Numer. Math. Appl. '84*, Bulgarian Academy of Sciences, Sofia, 1985.

[64] Maess, G. *Vorlesungen über Numerische Mathematik II*, Akademie Verlag, Berlin, 1988.

[65] Maus, A. "Delaunay triangulation and the convex hull of n points in expected linear time," *BIT* 24, 151–163 (1984).

[66] Mettke, H. "Biquadratische Interpolations- und Volumenausgleichssplines," *Beitr. Numer. Math.* 9, 135–145 (1981).

[67] Montefusco, L. B., and Casciola, G. "Algorithm 677: C^1 surface interpolation," *ACM TOMS* 15, 365–374 (1989).

[68] Moravec, O. "Lokale zweidimensionale C^1−Spline-Interpolation über einem Dreiecksgitter," Diplomarbeit, Universität Oldenburg, 1990. (This work is available from the library of the University of Oldenburg via inter-library loan.)

[69] Nielson, G. "Minimum norm interpolation in triangles," *SIAM J. Numer. Anal.* 17, 44–62 (1980).

[70] Nielson, G. M., and Franke, R. "Surface construction based upon triangulations." In Barnhill, R. E., and Boehm, W. (Eds.), *Surfaces in CAGD*, North Holland 1983.

[71] Nielson, G. M., and Franke, R. "A method for construction of surfaces under tension," *Rocky Mountain J. Math.* 14, 203–221 (1984).

[72] Nielson, G. M. "Coordinate free scattered data interpolation." In Chui, C. K., Schumaker, L. L., and Utreras, F. I. (Eds.), *Topics in Multivariate Approximation*, Academic Press, Orlando Fla., 1987.

[73] Percell, P. "On cubic and quartic Clough-Tocher finite elements," *SIAM J. Numer. Anal.* 13, 100–103 (1976).

[74] Pfluger, P. R., and Gmelig Meyling, R. H. J. "An algorithm for smooth interpolation to scattered data in R^2." In Lyche, T., and Schumaker, L. L. *Mathematical methods in Computer Aided Geometric Design*, Academic Press, Boston, 1989.

[75] Pilcher, D. "Smooth parametric surfaces." In Barnhill, R. E., and Riesenfeld, R. F. *Computer Aided Geometric Design*, Academic Press 1974.

[76] Poirier, D. J. "On the use of bilinear splines in economics," *J. of Econometrics* 3, 23–34 (1975).

[77] Powell, M. J. D. "Piecewise quadratic surface fitting for contour plotting." In: Evans, D., J. (Ed.), *Software for Numerical Mathematics*, Academic Press 1974.

[78] Powell, M. J. D., and Sabin, M. A. "Piecewise quadratic approximations on triangles," *ACM TOMS* 4, 316–325 (1977).

[79] Quak, E., and Schumaker, L. L. "C^1 surface fitting using data dependent triangulations." In Chui, C. K., Schumaker, L. L., and Ward, J. D. (Eds.), *Approximation Theory VI*, 1–4, Academic Press, Boston, 1989.

[80] Quak, E., and Schumaker, L. L. "Cubic spline fitting using data dependent triangulations," to appear in CAGD.

[81] Renka, R. J. "A triangle-based C^1 interpolation method," *ORNL/CSD - 103*, Oak Ridge National Laboratory, May 1982.

[82] Renka, R. J. "Interpolation of data on the surface of a sphere," *ACM Trans. Math. Softw.* 10, 417–436 (1984).

[83] Renka, R. J. "Algorithm 623: Interpolation on the surface of a sphere," *ACM Trans. Math. Softw.* 10, 437–439 (1984).

[84] Renka, R. J. "Algorithm 624: Triangulation and interpolation at arbitrarily distributed points in the plane," *ACM Trans. Math. Softw.* 10, 440–442 (1984).

[85] Renka, R. J., and Cline, A. K. "A triangle-based C^1 interpolation method," *Rocky Mountain J. Math.* 14, 223–237 (1984).

[86] Renka, R. J. "Multivariate interpolation of large sets of scattered data," *ACM TOMS* 14, 139–148 (1988).

[87] Renka, R. J. "Algorithm 660: QSHEP2D: Quadratic Shepard method for bivariate interpolation of scattered data," *ACM TOMS* 14, 149–150 (1988).

[88] Renka, R. J. "Algorithm 661: QSHEP3D: Quadratic Shepard method for trivariate interpolation of scattered data," *ACM TOMS* 14, 151–152 (1988).

[89] Sakai, M., and Usmani, R., A. "Biquadratic spline approximations," *Publ. RIMS, Kyoto Univ.* *20*, 431–446 (1984).

[90] Schmidt, F. "Positive 1D- und 2D-Interpolation mit Splines," Diplomarbeit, TU Dresden, 1988.

[91] Schmidt, J. W. "Results and problems in shape preserving interpolation and approximation with polynomial splines." In Schmidt, J. W., and Späth, H., (Eds.), *Splines in Numerical Analysis*, Akademie Verlag, Berlin, 1989.

[92] Schmidt, R. "Eine Methode zur Konstruktion von C^1−Flächen zur Interpolation unregelmäßig verteilter Daten." In Schempp, W., and Zeller, K. (Eds.), *Multivariate Approximation Theory II*, Birkhäuser, Basel, 1982.

[93] Schumaker, L. L. "Fitting surfaces to scattered data." In Lorentz, G. G., and Schumaker, L. L. (Eds.), *Approximation Theory II*, Academic Press 1976.

[94] Schumaker, L. L. "Numerical aspects of spaces of piecewise polynomials on triangulations." In Mason, J. C., and Cox, M. G. (Eds.), *Algorithms for Approximation*, Clarendon Press, Oxford, 1987.

[95] Schumaker, L. L. "Triangulation methods." In Chui, C. K., Schumaker, L. L, and Utreras, F. (Eds.), *Topics in Multivariate Approximation*, Academic Press, Orlando Fla., 1987.

[96] Seitelman, L. H. "New user-transparent edge conditions for bicubic spline surface fitting," *Rocky Mountain J. Math.* 14, 351–371 (1984).

[97] Shepard, D. "A two-dimensional interpolation function for irregularly spaced data," *Proc. 1968 ACM Nat. Conf.*, 517–524.

[98] Späth, H. "Zweidimensionale glatte Interpolation," *Computing* 4, 178–182 (1969). Errata: *Computing* 8, 200–201 (1971).

[99] Späth, H., "Two-dimensional exponential splines," *Computing* 7, 364–369 (1971).

[100] Späth, H. *One Dimensional Spline Interpolation Algorithms*, A K Peters, Boston, 1995.

[101] Späth, H. *Spline-Algorithmen zur Konstruktion glatter Kurven und Flächen*, 4th Edition, R. Oldenbourg-Verlag, München, 1986.

[102] Stark, A. "Volumentreue zweidimensionale Spline-Interpolation: Konstruktion und numerische Verfahren." Diplomarbeit, Universität Oldenburg, 1989. (This work can be obtained from the library of the University of Oldenburg via inter-library loan.)

[103] Stead, S. E. "Estimation of gradients from scattered data," *Rocky Mountain J. Math.* 14, 265–279 (1984).

[104] Tobler, W., Lau, J. "Interpolation of images via histosplines," *Comp. Graphics Image Proc.* 9, 77–81 (1979).

[105] Utreras, F. I. "Constrained surface construction." In Chui, C. K., Schumaker, L. L., and Utreras, F. I. (Eds.), *Topics in Multivariate Approximation*, 233–254, Academic Press, Orlando Fla., 1987.

[106] Vasil'ev, M. G., and Yufurev, V. S. "Bicubic spline interpolation in polar coordinates," *Zh. vychisl. Mat. mat. Fiz.* 18, 1600–1602 (1978).

[107] Young, J. D. "An optimal bicubic spline on a rectilinear mesh over a rectangle," *The Logistics and Transportation Review* 8, 33–40 (1972).

Index

Akima 95, 224, 275

B-splines 31, 74
barycentric coordinates 201,
    251
Bernstein polynomial 107
bicubic
    Hermite interpolation 93
    interpolation 71
    parametric interpolation 91
    shape-preservation 106
bilinear 13, 14
bimonotone 106, 126
biquadratic
    histosplines 108
    interpolation
      knots differ from nodes 49
      knots sames as nodes 31
    shape-preservation 68
birational
    Hermite interpolation 126
    histosplines 126
    interpolation 121
    shape-preservation 121

cardinal function 4, 5
collinear 163, 201
COND 173, 174, 176, 194
connection matrix 29, 33, 51, 72,
    122
contour line 234, 273

extrapolation 213, 251, 273

grid line 4, 5, 215
grid point 3, 4, 6

Hardy's multiquadrics 173, 216
histointerpolation 113
histospline 108, 119
Horner's rule 37, 54, 125

interpolating plane 12
interpolation
    positive 107
    convex 107
interpolation method
    local 68, 93, 95, 164
    global 13, 32, 68
    semi-global 95
interpolation problem 4

Klucewicz's method 224
knot point 6

Lagrange interpolant 5
lattice 4

natural bicubic spline 74

periodicity 74
planar equation 12
polar grid 283
product 4
product interpolation 31, 49

radial basis function 173
rectangle 3
rectangular grid 3, 4, 13

semi-C^1 103
shape-preservation 6, 68, 106, 125, 126, 283
Shepard's method 165, 215, 273
subrectangle 3
subtriangle 229, 232, 233, 249–251, 275

tension parameter 126
triangulation 195, 215, 249
    data-dependent 200, 273
    Delaunay 197, 202, 233, 273
    Thiessen 197
twist values 98, 103

volume heights 108, 110

Milton Keynes UK
Ingram Content Group UK Ltd.
UKHW020021071024
449327UK00032B/2881